Protein C
Biochemical and Medical Aspects

Protein C

Biochemical and Medical Aspects

Proceedings of the
International Workshop
Titisee, Federal Republic of Germany
July 9-11, 1984

Editor Irene Witt

W DE G Walter de Gruyter
Berlin · New York 1985

Editor

Irene Witt, Prof. Dr.
Klinisch-Chemisches Labor
Biochemisches Labor
Universitäts-Kinderklinik
Mathildenstraße 1
D-7800 Freiburg
Federal Republic of Germany

Library of Congress Cataloging in Publication Data

Protein C: biochemical and medical aspects.
Includes bibliographies and index.
1. Protein C-Congresses. I. Witt, Irene.
[DNLM: 1. C-Reactive Protein-congresses. WH 400 P967 1984]
QP93.5.P766 1985 612'.015752 85-16126
ISBN 0-89925-094-7 (U.S.)

CIP-Kurztitelaufnahme der Deutschen Bibliothek

Protein C: biochem. and med. aspects;
proceedings of the internat. workshop, Titisee,
Fed. Republic of Germany, July 9-11, 1984/ed. Irene Witt. -
Berlin; New York: de Gruyter, 1985.
ISBN 3-11-010222-6 (Berlin ...)
ISBN 0-89925-094-7 (New York)
NE: Witt, Irene [Hrsg.]

ISBN 3 11 010222 6 Walter de Gruyter · Berlin · New York
ISBN 0-89925-094-7 Walter de Gruyter, Inc., New York

 Printing: Gerike GmbH, Berlin. - Binding: Dieter Mikolai, Berlin. - Printed in Germany.

Preface

Thrombosis has been known since about 2 000 B.C. and pulmonary embolism was first described in 1676 by WISEMAN. In the 1840's Rudolf VIRCHOW recognized the connection between thrombosis and pulmonary embolism.

According to VIRCHOW's triad changes of the vessel wall, blood flow and blood properties are the most important pathogenic factors for the development of thrombi with the risk of consecutive embolism.

For a long time it was assumed that hypercoagulability is mainly caused by an increased intravascular thrombin generation. During the last decade a key role in the development of hypercoagulability was attributed to proteinase inhibitors. The importance of antithrombin III became well established when it was observed at the end of the 1960's that an inherited or acquired deficiency is combined with a high risk of thromboembolism.

A few years ago, it became evident that thrombin generation can be inhibited not only by proteinase inhibitors but also by the action of a specific proteinase - Protein C. The inhibition of coagulation by Protein C may be of equal importance to that of antithrombin III.

The protein, which has already been known since 1960 as autoprothrombin II a, had a rennaissance when in 1981 GRIFFIN found the first family with a hereditary deficiency of this protein and its association with a high risk of thromboembolic diseases or thromboembolic complications.

Since then there has been increasing interest and research regarding Protein C. It belongs to a complex system of interacting proteins, it needs vitamin K for the synthesis of its biologically active form, it needs a vitamin K dependent protein as cofactor and gains its full activity only after activation by thrombin and thrombomodulin. Moreover a specific inhibitor exists for Protein C.

It is imposing that only a few years were necessary to clarify the complex basic biochemistry of Protein C and related proteins. New insights into the regulation of coagulation by plasmatic inhibitors arose. Many families have since been found with a congenital deficiency of this protein and many cases of spontaneous or recurrent thrombosis could be elucidated as a deficiency of Protein C or its cofactor Protein S.

Despite advanced knowledge there are still many problems in determining Protein C functionally. Most of them arise from its characeristic features. The difficulties of the functional assay in routine use are presumably the reason why little information is available about the changes in Protein C activity during the course of various diseases.

The aim of the workshop on Protein C, which was held form July 9th - 11th, 1984, in Titisee near Freiburg i.Br., was to sum up present knowledge in the biochemistry, physiology and pathophysiology of the inhibitor and related proteins. Important topics were the possibilities of determining Protein C concentration immunologically or the functional activity by chromogenic or clotting assays.

The basic information on Protein C and connected proteins was followed by a number of clinical reports on inherited and acquired disorders. Multifarious open questions were outlined in the lectures and subsequent discussions.

I hope the proceedings of the workshop which are documented in this book will reflect the current status of the complex Protein C system and spread information on this new field of cagulation.

I wish to express special thanks to Dr. Ernst Zimmer, Mannheim, for his thorough help with the editing of the manuscripts. Sincere gratitude is extended to Mrs. Monika Krumnow, Mrs. Evelyne Glowka and Dr. Rudolf Weber of Walter de Gruyter Publishers, whose helpful cooperation has made the task of editing this book a pleasant one.

Freiburg i. Br., July 1985 Irene Witt

Acknowledgement

The workshop was made possible by the generous sponsorship of Boehringer Mannheim GmbH. Further valuable contributions came from MERZ & DADE (AHS), München, Deutsche KABI VITRUM, München, PENTAPHARM, Basel and SCHWAB, Brunn (Austria). I wish to express my sincere gratitude to all of them.

LIST OF PARTICIPANTS

Dr. J. Amiral	Diagnostica Stago 6 ter rue Denis Papin 92600 Asnieres
Prof. Dr. E. A. Beck	Hämatologisches Zentrallabor Inselspital Bern Freiburgstr. 30 CH-3013 Bern
Prof. Dr. H. Beeser	Institut f. Transfusionsmedizin der Universität Freiburg Hugstetterstr. 55 D-7800 Freiburg i.Br.
Dr. R. M. Bertina	Hemostasis and Thrombosis Research Unit Academisch Ziekenhuis Leiden Rijnsburgerweg 10 2333 AA Leiden
Dr. W. Brockhaus	Klinikum Nürnberg Flurstr. 17 D-8500 Nürnberg
Dr. A. W. Broekmans	Hemostasis and Thrombosis Research Unit Academisch Ziekenhuis Leiden Rijnsburgerweg 10 2333 AA Leiden
Dr. J. P. Cazenave	Service d'Hemostase et de Thrombose Centre le Transfusion Sanguin F-67100 Strasbourg
Dr. M. Colucci	Center for Thrombosis and Vascular Research Katholieke Universiteit Leuven Department of Medical Research B-3000 Leuven
Dr. H. Droste	Boehringer Mannheim GmbH Sandhoferstr. 116 D-6800 Mannheim

Dr. J. H. Griffin	Dept. of Molecular Immunology Scripps Clinic and Research Foundation 10666 North Torrey Pines Road La Jolla, California 92037
Prof. Dr. W. Heller	Chirurgische Univ.-Klinik Calwer Straße 7 D-7400 Tübingen
Dr. P. Hellstern	Universitätskliniken Abtlg. Klin. Haemostaseologie und Transfusionsmedizin D-6650 Homburg/Saar
Dr. H. P. Kreuzer	Biochemisches Labor Universitäts-Kinderklinik Mathildenstraße 1 D-7800 Freiburg i. Br.
Prof. Dr. K. Lechner	I. Medizinische Univ.-Klinik Lazarettgasse 14 A-1090 Wien
Dr. H. Lill	Boehringer Mannheim GmbH Biochemica Werk Tutzing Bahnhofstr. 5 D-8132 Tutzing/Obb.
Prof. Dr. G. Müller-Berghaus	Klinische Forschungsgruppe für Blutgerinnung und Thrombose Max-Planck-Gesellschaft zur Förderung der Wissenschaften e.V. Gaffkystr. 11 D-6300 Gießen
Doz. Dr. G. Oehler	Klinikum der Justus-Liebig-Univ. Zentrum für Innere Medizin Klinikstr. 36 D-6300 Gießen
Prof. Dr. W. G. Owen	Cardiovascular Center, Departments of Pathology and Biochemistry University of Iowa Iowa City, Iowa 52242
Dr. I. Pabinger-Fasching	I. Medizinische Univ.-Klinik Lazarettgasse 14 A-1090 Wien
Dr. F. W. Rabe	Deutsche Kabi Vitrum GmbH Levelingstr. 18 D-8000 München 80

Prof. Dr. L. Róka	Klinikum der Justus-Liebig-Univ. Inst. f. Klinische Chemie und Pathobiochemie Klinikstr. 32 b D-6300 Gießen
Dr. S. Rosén	AB KABI Peptide Research S-43122 Mölndal
Prof. Dr. I. Scharrer	Zentrum der Inneren Medizin der Univ.-Kliniken Frankfurt Theodor-Stern-Kai 7 D-6000 Frankfurt/M.
Doz. Dr. U. Schmitz-Huebner	Medizinische Klinik u. Poliklinik der Westfälischen Wilhelms-Univ. Abtlg. Innere Medizin A Domagkstr. 3 D-4400 Münster
Prof. Dr. T. H. Schöndorf	Klinikum der Justus-Liebig-Univ. Zentrum f. Innere Medizin Klinikstr. 36 D-6300 Gießen
Dr. H. H. Seydewitz	Biochemisches Labor Universitäts-Kinderklinik Mathildenstr. 1 D-7800 Freiburg i.Br.
Dr. E. Spanuth	Boehringer Mannheim GmbH Sandhoferstr. 116 D-6800 Mannheim
Dr. T. Sugo	Department of Clinical Chemistry University of Lund Malmö General Hospital S-21401 Malmö
Prof. Dr. K. Suzuki	Dept. of Laboratory Medicine Mie University School of Medicine Tsu-City, Mie 514
Prof. Dr. T. Vukovich	Institut f. Med. Physiologie Schwarzspanierstr. 17 A-1090 Wien
Prof. Dr. F. J. Walker	Terre Haute Center for Medical Education at Indiana State Univ. 135 Holmstedt Hall Terre Haute, Indiana 47809

Prof. Dr. H. Wehinger
Städtische Kliniken Kassel
Kinderklinik
Mönchebergstr. 41/43
D-3500 Kassel

Prof. Dr. E. Wenzel
Universitätskliniken
Abtlg. Klin. Haemostaseologie
und Transfusionsmedizin
D-6650 Homburg/Saar

Prof. Dr. I. Witt
Biochemisches Labor
Universitäts-Kinderklinik
Mathildenstr. 1
D-7800 Freiburg i.Br.

CONTENTS

BIOCHEMISTRY AND PHYSIOLOGY OF PROTEIN C

Whyte G. Owen
University of Iowa, Iowa City, Iowa 52242, USA

Identified originally as an anticoagulant activity elicited by treating prothrombin preparations with thrombin (1), protein C was purified as a by-product of prothrombin purification (2), after having been identified as a third, unknown, component in an ion-exchange chromatogram (hence, "C"). As a member of the family of vitamin K-dependent zymogens of serine proteinases, protein C is γ-carboxylated at the first eleven glutamate residues (3, 4), and has the core structure of the chymotrypsin family. The zymogen is comprised of two non-identical polypeptides, of 155 and 260 amino acid residues, linked by a single disulfide. Of yet unknown significance, aspartate-71 is ß-hydroxylated (5). Removal of an N-terminal tetradecapeptide from the heavy chain activates protein C to an arginine-specific endopeptidase, with the catalytic center confined to the heavy chain.

The only natural substrates of activated protein C so far identified are the active forms of factors V and VIII (6-8). In a reaction dependent on phospholipid (or platelets), factors Va and VIIIa are inactivated rapidly by catalytic concentrations of activated protein C, owing to cleavage of specific peptide bonds. Cleavage of procoagulant cofactors appears to account for the anticoagulant activity of activated protein C, which is comparable *in vivo* and *in vitro* in blood or plasma (9, 10). The anticoagulant activity of activated protein C is species-specific, a finding that has led to identification of protein S, discussed elsewhere in this volume, as an obligate cofactor for activated protein C (11-13).

Protein C - Biochemical and Medical Aspects

A fibrinolytic action of activated protein C has been observed in dogs (9, 10, 14). Intravascular activated protein C elicits a transient but profound fibrinolytic state that arises from circulation of a tissue-type plasminogen activator. The response requires both humoral and cellular elements, and is likely to proceed via multiple steps. However, as discussed in detail elsewhere in this volume, the response is not observed in squirrel monkeys given anticoagulant doses of human or endogenous (monkey) activated protein C (15). Whether the response in dogs is unique or much more profound than in other species has not been ascertained.

Although protein C is activated slowly by thrombin, rapid activation of the zymogen, like that of other vitamin K-dependent zymogens requires a cofactor. Termed thrombomodulin, the cofactor was identified in perfused heart and cultured endothelium preparations (16, 17) and was purified from detergent extracts of rabbit lung (18). The purified protein is a single polypeptide having a molecular weight of 75,000, contains about 10 % proline, and 38 half-cystines. Seemingly unique to vascular endothelium, thrombomodulin is an integral membrane protein that binds thrombin stoichiometrically with high affinity (K_D = 0.5 nM) and specificity.

The bound enzyme is a potent activator of protein C, but is inactive toward other biological (procoagulant) substrates (19, 20), heparin cofactor II and α_2-macroglobulin. This change of specificity is reversible upon dissociation of thrombin from thrombomodulin. Studies of the Ca^{2+}-dependence of activation of native and modified derivatives of protein C indicate that membrane components other than thrombomodulin also participate in thrombin-catalzyed protein C activation (21, 22).

From the viewpoint of biochemistry, protein C fits well into the family of cofactor-regulated vitamin K-dependent zymogens. From the viewpoint of physiology, we may expect that protein C and thrombomodulin are but two members of a mostly unknown regulatory system that should come to be as complex as the coagulant system. The pathology of protein C has already indicated that participation of protein C in the strict control of procoagulant expression is requisite to homeostasis.

References

1. Mammen, E.F., Thomas, W.R., Seegers, W.H.: Thromb. Haemostas. 5, 218-249 (1960).
2. Stenflo, J.: J. Biol. Chem. 251, 355-363 (1976).
3. Fernlund, P., Stenflo, J.: J. Biol. Chem. 257, 12170-12179 (1982).
4. Stenflo, J., Fernlund, P.: J. Biol. Chem. 257, 12180-12190 (1982).
5. Drakenberg, T., Fernlund, P., Roepstorff, P., Stenflo, J.: Proc. Natl. Acad. Sci. USA 80, 1802-1806 (1983).
6. Kisiel, W., Canfield, W.M., Ericsson, L.H., Davie, E.W.: Biochemistry 16, 5824-5831 (1977).
7. Vehar, G.A., Davie, E.W.: Biochemistry 19, 401-410 (1980).
8. Walker, F.J., Sexton, P.W., Esmon, C.T.: Biochim. Biophys. Acta. 571, 333-342 (1979).
9. Comp, P.C., Esmon, C.T.: J. Clin. Invest. 68, 1221-1228 (1981).
10. Comp. P.C., Jacocks, R.M., Ferrell, G.L., Esmon, C.T.: J. Clin. Invest. 70, 127-134 (1982).
11. Walker, F.J.: J. Biol. Chem. 255, 5521-5524 (1980).
12. Walker, F.J.: Thromb. Res. 22, 321-327 (1981).
13. Walker, F.J.: J. Biol. Chem. 256, 11128-11131 (1981).
14. Seegers, W.H., McCoy, L.E., Groben, H.D., Sakuragawa, N., Agrawal, B.B.L.: Thromb. Res. 1, 443-460 (1972).
15. Colucci, M., Stassen, J.M., Collen, D.: J. Clin. Invest. 74, 200-204 (1984).

16. Esmon, C.T., Owen, W.G.: Proc. Natl. Acad. Sci. USA 78, 2249-2252 (1981).

17. Owen, W.G., Esmon, C.T.: J. Biol. Chem. 256, 5532-5535 (1981).

18. Esmon, N.L., Owen, W.G., Esmon, C.T.: J. Biol. Chem. 257, 859-864 (1982).

19. Esmon, C.T., Esmon, N.L., Harris, K.W.: J. Biol. Chem. 257, 7944-7947 (1982).

20. Esmon, N.L., Carroll, R.C., Esmon, C.T.: J. Biol. Chem. 258, 12238-12242 (1983).

21. Esmon, N.L., DeBault, L.E., Esmon, C.T.: J. Biol. Chem. 258, 5548-5553 (1983).

22. Johnson, A.E., Esmon, N.L., Laue, T.M., Esmon, C.T.: J. Biol. Chem. 258, 5554-5560 (1983).

MECHANISM OF ACTION OF PROTEIN C

Hans Peter Schwarz and John H. Griffin

Department of Immunology, Scripps Clinic and Research Foundation, La Jolla, California 92037, USA

Introduction

Human protein C, a vitamin K-dependent glycoprotein consisting of disulfide-linked heavy and light polypeptide chains, circulates as an inactive zymogen in plasma at a concentration of 4 µg/ml (1). Recent data suggests that protein C is initially synthesized in the liver as a single chain molecule that undergoes processing into a two chain molecule by proteolytic cleavage (2). The heavy chain constitutes the serine protease part of the molecule whereas the vitamin K-dependent structure is located in the NH_2-terminal part of the light chain containing 11 γ-carboxyglutamic acid residues. During activation of human protein C by thrombin, the Arg_{12}-Leu_{13} bond is cleaved at the amino-terminal end of the heavy chain and a dodecapeptide of 1,400 molecular weight is released (3).
This protein C activation peptide is measurable in humans, reflecting the degree of protein C activation in the vascular system (4). The rate of the activation of protein C by thrombin, the only known physiological activator, is catalysed several thousand fold after thrombin complexes with thrombomodulin, an endothelial cell cofactor (5). Human thrombomodulin has been purified recently from placenta and some striking differences between human placenta thrombomodulin and rabbit lung thrombomodulin have been reported (6). Human thrombomodulin bound to thrombin does not neutralize the pro-

Protein C - Biochemical and Medical Aspects

coagulant properties of thrombin in contrast to rabbit thrombomodulin (Majerus et al., American Heart Association Meetings, Miami, 1984).

Human protein C in its activated form has powerful antithrombotic properties by virtue of at least three effects on hemostasis. Protein C is an anticoagulant by inactivating the coagulant cofactors, factors Va and VIIIa, that mediate activation of factor X by factor IXa and activation of prothrombin by factor Xa (7, 8, 9, 10).

Activated protein C (APC) destroys platelet prothrombin-converting activity by inhibiting the binding of factor Xa to platelets through inactivation of platelet factor Va, the Xa receptor, thus resulting in a reduction of thrombin formation (11, 12).

Activated protein C has fibrinolytic properties both _in vivo_ and _in vitro_ (13, 14). The fibrinolytic potential of cultured endothelial cells is increased by APC and an inhibitor of plasminogen activator (antiactivator) is neutralized by APC in _in vitro_ blood clot lysis assays (15, Sakata, Loskutoff and Griffin, unpublished results). The intravenous infusion of bovine APC into dogs apparently elevates circulating plasminogen activator (13).

Protein C as an anticoagulant

In 1960 Mammen et al. detected a thrombin-dependent anticoagulant in a preparation of bovine vitamin K-dependent proteins (16, 17). After purification of this anticoagulant, named by these authors autoprothrombin II-A, the identity of APC to it was demonstrated (18). In spite of initial suggestions that APC functions as an competitive inhibitor of prothrombin activation (17, 18), it is clear now that the anticoagulant activity is mediated through the selective proteolytic inactivation of factor VIII and factor V.

Infusion of thrombin (1 U/kg/min) into dogs resulted in the generation of anticoagulant activity within 5 min of starting the infusion as shown by prolonged partial thromboplastin times and factor Xa clotting times (5). The nature of the anticoagulant was identified as APC by partial purification of the anticoagulant from the dog plasma and by the observation that antibodies to APC added to the plasma before measuring the clotting time abolished the anticoagulant effect (5). In retrospect, the anticoagulant effect of APC was also probably demonstrated during other experiments in which low levels of thrombin were infused into animals resulting in the appearance of anticoagulant activity (19).

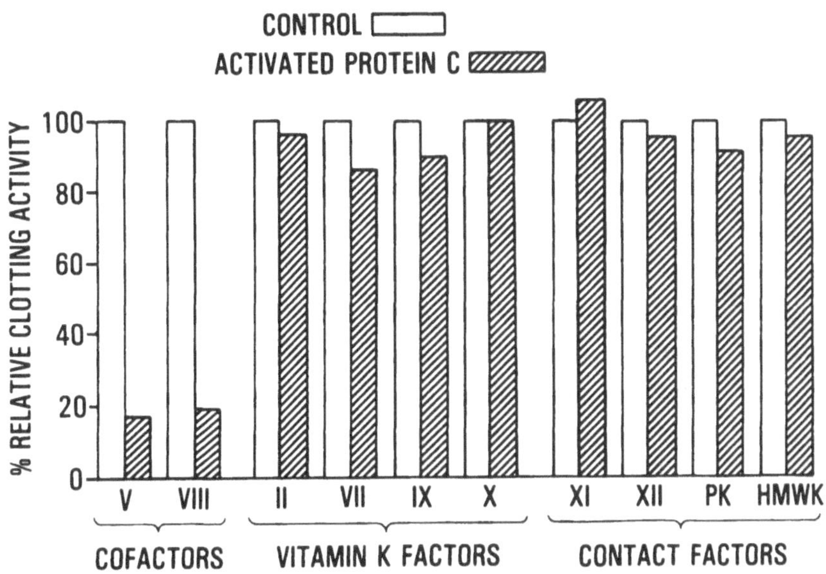

Fig. 1. Effect of activated protein C on coagulation factors in normal plasma. Activated protein C (3.0 µg/ml final) was combined with citrated normal human plasma (500 µl) in the presence of cephalin (58 µg/ml final) and calcium ions (4 mM final) in a total volume of 600 µl. After 3 min at 37°C, EDTA or purified antiprotein C antibodies were added, and dilutions of the mixture were assayed for individual coagulation factors. The height of each bar represents percent clotting activity compared to the control (open bars). Plasma plus activated protein C is represented by bars with diagonal lines. From Marlar, Kleiss and Griffin (20).

It was shown by Marlar et al. that human APC in the presence of phospholipids and calcium ions significantly prolongs the prothrombin time, the activated partial thromboplastin time and the partial thromboplastin time. No effect on thrombin time by APC was observed (20). As seen in Fig. 1, the anticoagulant activity of APC in plasma is due to the decrease of the clotting activity of factor V and factor VIII by APC since greater than 80 % of each activity were lost in a 3 minute incubation of APC with normal plasma in the presence of phospholipids and calcium ions (20). The inactivation of factors V and VIII in plasma due to APC showed a dose response relationship between the proteolytic destruction of factors V and VIII and the amount of APC present in the incubation mixture (20).

The inactivation of the activated form of factor VIII:C, factor VIIIa, and of factor Va by APC is much more rapid than that of the nonactivated forms. APC loses all anticoagulant activity when the active site is blocked with diisopropyl fluorophosphate (20). The contact system proteins, other vitamin K-dependent coagulation proteins, and fibrinogen are not affected by APC.

When the molecular changes during the inactivation of factor VIII and V by APC were studied, the polypeptide cleavage pattern of factor VIII:C caused by APC showed similarity to the proteolytic cleavage pattern of factor V (21).

During the inactivation reaction of purified human factor VIII:C by APC, a diminution of all factor VIII:C polypeptides of molecular weight between 92,000 and 188,000, leaving intact a doublet of molecular weight 79,000 to 80,000, occurs as shown on SDS polyacrylamide gel electrophoresis (21) (Fig. 2). The generation of a new major polypeptide of molecular weight 45,000 appears during this APC-induced proteolytic process. A time course study of the inactivation of purified factor VIII:C by APC revealed a progressive disappearance of factor VIII:C high molecular weight polypeptides as factor VIII activity decreased while the new peptide of molecular weight 45,000 appeared to increase in concentration (Fig. 2). Both human and bovine factor Va are molecules consisting of heavy and light chains both required for activity and held together non-covalently through Ca^{++}-dependent interactions. The inactivation of thrombin activated human factor Va is associated with a rapid cleavage of the 105,000 molecular weight heavy chain accompanied by a reduction in factor V activity (22 - 24). The degradation products of the heavy chain have molecular weights of 21,000 to 70,000. The light chain of factor Va is slowly degraded only upon prolonged incubation with APC. The degradation of the heavy chain of factor Va by APC results in a decrease in the affinity of factor Va for prothrombin with a successive loss of the ability to bind prothrombin (23). Factor Va binds to phospholipids or platelets via the light chain. Degradation of platelet-bound factor Va by APC results in the same cleavages of the heavy chain as when phospholipid-bound factor Va is cleaved. The APC-dependent proteolysis of platelet-bound factor Va, which is part of the factor Xa receptor, can be inhibited by factor Xa, thus emphasizing several potential roles of factor Xa in the regulation of the activation of prothrombin (22 - 25).

Protein C and protein S deficiency and thrombotic disease

The first evidence for the _in vivo_ role of protein C and for its clinical importance as a potent antithrombotic agent came from this lab when the first protein C deficient family presenting with a long and outstanding history of recurrent venous thromboembolism was described in 1981 (26). The propositus of the familiy had 41 % of protein C antigen and the father and uncle had 35 % and 45 % of normal protein C levels, respectively. All these men had recurrent venous thrombotic disease. Based on these results, our lab proposed that an inherited deficiency of protein C in heterozygote patients who exhibit approximately half normal plasma levels

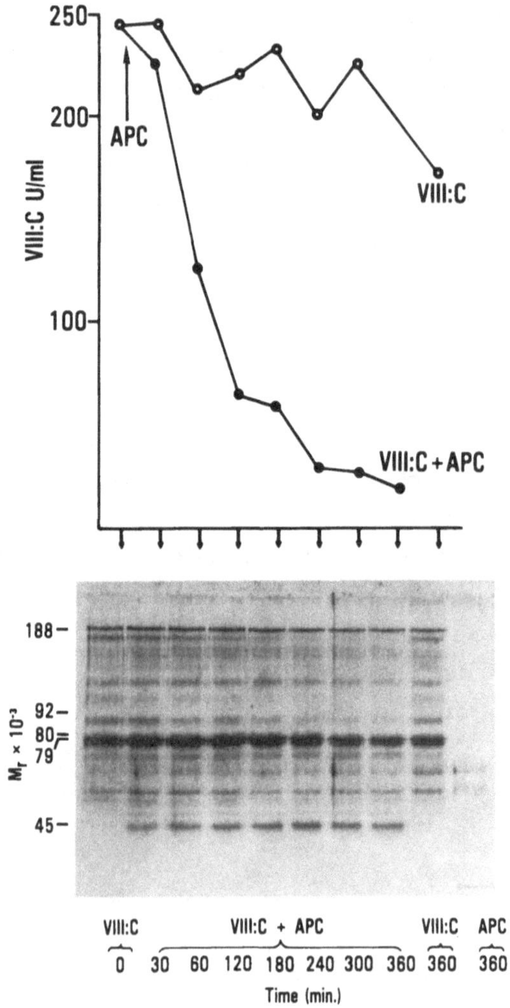

Fig. 2. Inactivation of factor VIII:C activity by APC (upper panel) with reduced SDS gel pattern factor VIII:C (lower panel). Gel lanes (~20 µg VIII:C) contained: control (VIII:C 0 min); VIII:C at various times after addition of APC (30-360 min); control (VIII:C 360 min); and APC alone (APC 360 min). In the activity plot the point at which APC was added is indicated with an arrow. From Fulcher, Gardiner, Griffin and Zimmerman (21).

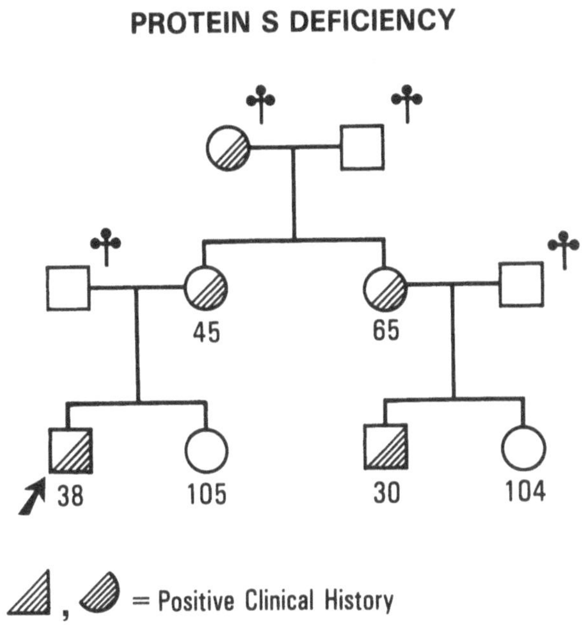

Fig. 3. Family tree of protein S-deficient family. The arrow denotes the propositus. Numbers indicate the percentage of protein S antigen. ○ , Females; □, Males; †, deseased family members, ◍ , ▨, positive clinical history. From Schwarz et al. (40).

to <u>in vitro</u> blood clot lysis assays accelerates lysis of clots and a requirement of cells for the fibrinolytic effect of APC was suggested (13).

When the effect of bovine APC on the fibrinolytic activity of cultured bovine aortic endothelial cells was investigated in our lab, a rapid, dose-dependent increase of fibrinolytic activity both in the conditioned media and in cellular extracts was observed (15). The increased fibrinolytic activity appeared to reflect increases in both urokinase and tissue-type plasminogen activators produced by these cells. The treatment of endothelial cells with APC results also in a rapid dose-dependent decrease in a fibrinolytic inhibitor, of protein C antigen can be associated with a tendency to venous thrombotic disease and that this disease is inherited as an autosomal trait (26). Since our initial finding in 1981, over 100 families have been identified between 1982 and 1984 with inherited protein C deficiency associated with thrombophilia (27). The observed low levels of protein C antigen in patients with disseminated intravascular coagulation (1, 28) are consistent with the hypothesis that protein C is a physiologic participant in the pathways of blood coagulation. Moreover, in patients with indications of liver disease (1, 28), protein C antigen levels are low, consistent with the assumption that protein C synthesis occurs in the liver. This was recently shown when Fair et al. demonstrated PC production in hepatoma cell lines (29). The relationship between purpura fulminans and protein C deficiency was first established by our lab (30 - 32). Homozygote protein C deficiency with an absence of detectable protein C presenting as purpura fulminans in the newborn is usually associated with death due to extreme thrombotic complications, indicating the vital role of protein C unless oral anticoagulant therapy or substitution therapy is given.

The rate of inactivation of factor Va by APC can be enhanced by protein S, another vitamin K-dependent plasma protein, by promoting the binding of APC to phospholipids (33). Equimolar binding of protein S to APC on the surface of phospholipids is required to form a factor Va inactivation complex. Protein S modified by thrombin, however, seems to inhibit the complex formation and the inactivation of factor Va by APC (33, 34). In plasma immunologically depleted of protein S, the normal anticoagulant activity of APC is not readily apparent and reconstitution of such a plasma with purified protein S restores the normal anticoagulant activity of APC (35). Our laboratory has shown that the inactivation of plasma factor VIII by APC is also stimulated by protein S and that the inactivation of purified human factor VIII:C by APC is enhanced by protein S in a dose-dependent manner (35).

Comp and Esmon and Dahlbäck and Stenflo presented evidence that APC inactivated the platelet associated factor V concomitant with a decrease in the binding of factor Xa to platelets (11, 12). Recent studies by Suzuki et al. have shown that protein S is essential for the efficient inactivation of platelet associated factor Va by APC (36). We recently found that protein S antigen is present in washed human platelets and is released upon thrombin stimulation, and we suggested that protein S may play a role as a negative regulator of prothrombin activation by enhancing the inactivation of platelet Va on platelets (37).

The very recent discovery of protein S deficiency in families (Fig. 3) and in unrelated individuals associated with venous thrombosis by Comp and Esmon (38, 39) and our lab (40) suggests the cofactor activity of protein S observed in vitro may also be important for the in vivo activity of protein C as an anticoagulant enzyme. The heterozygous deficiency of protein S appears to be similar to the heterozygous deficiency of protein C in that there is a remarkable history of recurring venous thrombotic disease often beginning in young adulthood (38 - 40). Fig. 3 shows the family tree of a typical protein S deficient family (40).

Protein C as a fibrinolytic agent

After the anticoagulant activity of APC was described, it was also shown that infusion of bovine APC into dogs increases the rate of lysis of whole blood clots _ex vivo_ due to an elevation of circulating lysine-Sepharose-adsorbable plasminogen activator (13, 17). Neither protein C zymogen nor APC treated with DFP was able to enhance clot lysis, suggesting that the observed response requires the active site of APC. Comp and Esmon suggested that APC generates a secondary messenger which in turn causes the elevation of circulating plasminogen activator activity (13). Other groups had shown that APC added designated antiactivator (15), that was originally described by Loskutoff et al. (41). This antiactivator, which is present in endothelial cells and platelets and released from platelets upon platelet stimulation, is capable of inhibiting both urokinase and tissue-type plasminogen activators (41, 42). Further studies (Sakata, Loskutoff and Griffin, unpublished data) have shown that in a radiolabelled clot lysis assay system containing purified tissue plasminogen activator, antiactivator, plasminogen, ^{125}I-fibrinogen and thrombin, the stimulation of fibrinolysis was caused by the addition of APC. In the absence of antiactivator, APC did not stimulate clot lysis. When clot lysis systems containing washed human platelets were used to study APC-dependent fibrinolysis, a marked decrease in antiactivator activity was detected by the reverse fibrin autography (Sakata, Loskutoff and Griffin, unpublished data). APC is also capable of decreasing antiactivator activity of whole blood and serum. Purified antiactivator rapidly neutralizes the enzymatic amidolytic activity of purified APC in a dose-dependent manner. Therefore, it appears that a major mechanism for protein C-dependent fibrinolysis in whole clot lysis assays involves neutralization of antiactivator by APC. The observed <u>in-vivo</u> fibrinolytic effect of APC may be caused by direct neutralization of antiactivator or by an increase in circulating plasminogen activators, or both. Whether the antithrombotic activity of APC depends on its anticoagulant or profibrinolytic qualities or both remains to be determined. Nonetheless, the discovery of protein S deficiency associated with thrombotic disease (38 - 40) similar to protein C deficiency suggests that the anticoagulant properties of APC:protein S complexes may be very important for the physiologic regulation of thrombosis.

The scheme in Fig. 4 summarizes the central role of protein C and protein S in the regulation of the blood coagulation pathways. In this scheme, APC and protein S function by inactivating factors Va and VIIIa by limited proteolysis. Thus, there is a thrombin-dependent feedback mechanism for turning off the blood coagulation pathways. The physiologically essential role of this protein C pathway is demonstrated by the discovery of protein C and protein S deficiency associated with venous thrombotic disease.

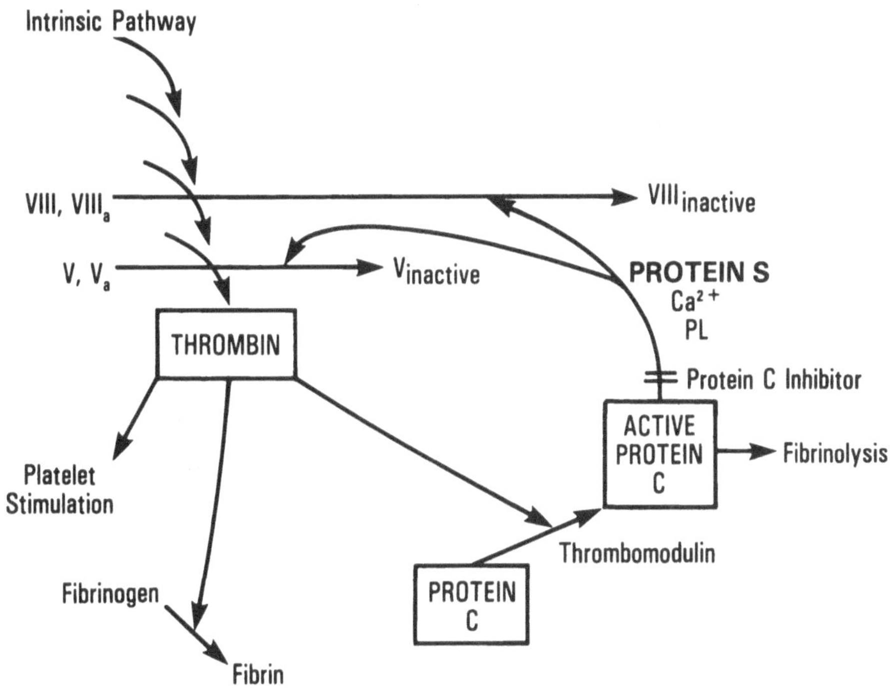

Fig. 4. Schematic representation of the protein C pathway in relation to the intrinsic blood coagulation pathway. See text for discussion.

References

1. Griffin, J.H., Mosher, D.F., Zimmerman, T.S., Kleiss, A.J.: Blood 60, 261-264 (1982).

2. Long, G.L., Belagaje, R.M., MacGillivray, R.T.A.: Proc. Natl. Acad. Sci. USA 81, 5653-5656 (1984).

3. Stenflo, J.: Semin. Thromb. Haemostas. 10 (2), 109-121 (1984).

4. Bauer, K.A., Kass, B.L., Beeler, D.L., Rosenberg, R.D.: J. Clin. Invest. 74, 2033-2041 (1984).

5. Comp, P.C., Jacocks, R.M., Ferrell, G.L., Esmon, C.T.: J. Clin. Invest. 70, 127-134 (1982).

6. Salem, H.H., Maruyama, I., Majerus, P.W.: J. Biol. Chem. 259, 12246-12251 (1984).

7. Kisiel, W., Ericsson, L.H., Davie, E.W.: Biochemistry 15, 4893-4900 (1976).

8. Kisiel, W., Davie, E.W.: Methods in Enzymology 80, 320-332 (1981).

9. Esmon, C., Comp, P., Walker, F.J.: In: Vitamin K Metabolism and Vitamin K-dependent Proteins. Suttie, J., ed., University Park Press, Baltimore, 72-83 (1980).

10. Vehar, G., Davie, E.W.: Biochemistry 19, 401-410 (1980).

11. Comp, P.C., Esmon, C.T.: Blood 54, 1272-1281 (1979).

12. Dahlbäck, B., Stenflo, J.: Eur. J. Biochem. 107, 331-335 (1980).

13. Comp, P.C., Esmon, C.T.: J. Clin. Invest. 68, 1221-1228 (1981).

14. Taylor, F.B., Jr., Caroll, R.C., Gerrard, J., Esmon, C.T., Radcliff, R.D.: Fed. Proc. 40(7), 2092-2098 (1981).

15. Sakata, Y., Curriden, S., Lawrence, D., Griffin, J.H., Loskutoff, D.J.: Proc. Natl. Acad. Sci. USA, in press (1985).

16. Mammen, E., Thomas, W., Seegers, W.H.: Thromb. Diath. Haemorrh. 5, 218-250 (1960).

17. Murano, G., Seegers, W.H., Zolton, R.P.: Thromb. Diath. Haemorrh. 57 (Suppl), 305-314 (1974).

18. Seegers, W.H., Novoa, E., Henry, R.L., Hassouna, H.I.: Thromb. Res. 8, 543-552 (1976).

19. Wetmore, R., Gurewich, V.: Scand. J. Haem. 12, 204-212 (1974).

20. Marlar, R.A., Kleiss, A.J., Griffin, J.H.: Blood 59, 1067-1072 (1982).

21. Fulcher, C.A., Gardiner, J.E., Griffin, J.H., Zimmerman, T.S.: Blood 63(2), 486-489 (1984).

22. Walker, F.J., Sexton, D.W., Esmon, C.T.: Biochim. Biophys. Acta 511, 333-342 (1979).

23. Guinto, E.R., Esmon, C.T.: J. Biol. Chem. 259, 13986-13992 (1984).

24. Suzuki, K., Stenflo, J., Dahlbäck, B., Teodorsson, B.: J. Biol. Chem. 258, 1914-1920 (1983).

25. van de Waart, P., Bruls, H., Hemker, C., Lindhout, T.: Biochemistry 22, 2427-2432 (1983).

26. Griffin, J.H., Evatt, B., Zimmerman, T.S., Kleiss, A.J., Wideman, C.: J. Clin. Invest. 68, 1370-1373 (1981).

27. Broekmans, A.W., Veltkamp, J.J., Bertina, R.: N. Engl. J. Med. 309, 340-344 (1983).

28. Mannucci, P.M.: Lancet, August 28, 463-467 (1982).

29. Fair, D.: Circulation 70 (4), 817 (1984) Abstract.

30. Branson, H.E., Katz, J., Marble, R., Griffin, J.H.: Lancet ii, 1165-1168 (1983).

31. Seligsohn, U., Berger, A., Abend, M., Rubin, L., Attias, D., Zivelin, A., Rapaport, S.I.: N. Engl. J. Med. 310, 559-562 (1984).

32. Sills, R.M., Marlar, R.A., Montgomery, R.R., Deshpande, G.N., Humbert, J.R.: J. Pediat. 105, 409-413 (1984).

33. Walker, F.J.: Semin. Thromb. Haemostas. 10 (2), 131-138 (1984).

34. Suzuki, K., Nishioka, J., Hashimoto, S.: J. Biol. Chem. 94, 699-705 (1983).

35. Gardiner, J.E., McGann, M.A., Berridge, C.W., Fulcher, C.A., Zimmerman, T.S., Griffin, J.H.: Circulation 70 (4), 820 (1984) Abstract.

36. Suzuki, K., Nishioka T., Matsuda, M., Murayama, H., Hashimoto, S.: J. Biochem. 96, 455-460 (1984).

37. Schwarz, H.P., Heeb, M.J., Griffin, J.H.: Circulation 70 (4), 821 (1984) Abstract.

38. Comp, P.C., Esmon, C.T.: N. Engl. J. Med. 311, 1525-1528 (1984).

39. Comp, P.C., Nixon, R.R., Cooper, R.M., Esmon, C.T.: J. Clin. Invest. 74, 2082-2088 (1984).

40. Schwarz, H.P., Fischer, M., Hopmeier, P., Batard, M.A., Griffin, J.H.: Blood 64, 1297-1300 (1984).

41. Loskutoff, D.J., Van Mourik, J.A., Erickson, L.A., Lawrence, D.: Proc. Natl. Acad. Sci. USA 80, 2956-2960 (1983).

42. Erickson, L.A., Ginsberg, M.H., Loskutoff, D.J.: J. Clin. Invest. 74, 1465-1477 (1984).

BIOCHEMISTRY OF PROTEIN S AND ITS ROLE IN THE REGULATION OF THE ANTICOAGULANT ACTIVITY OF ACTIVATED PROTEIN C

Frederick J. Walker
Indiana University School of Medicine, Terre Haute Center for Medical Education, Terre Haute, IN 47809, USA

I. Anticoagulant properties of activated protein C

Activated protein C is a serine protease that is derived from the vitamin K-dependent protein, protein C, and exhibits potent anticoagulant activity. The existence of a vitamin K-dependent protein that inhibited blood coagulation was first observed by Seegers (1), who observed that the treatment of a prothrombin preparation with thrombin yielded an inhibitor of coagulation. Seegers named this protein autoprothrombin IIa (1, 2, 3, 4). He observed that this protein, isolated from bovine plasma, was a potent inhibitor of coagulation and a stimulator of fibrinolysis in the dog (5). Following the examination of the effects of autoprothrombin IIa on prothrombin activation Seegers concluded that this protein was a competitive inhibitor of factor Xa in the prothrombinase complex (2, 6). An independent line of investigation was subsequently initiated by Stenflo, who had identified a new vitamin K-dependent protein in bovine plasma (7). It was named protein C. This protein was able to bind phospholipids in the presence of calcium and could be activated to a serine esterase by trypsin (8) or thrombin (9). Though at the time it was unclear, it is now known that autoprothrombin IIa and activated protein C are identical (4). Kisiel was first to report that activated protein C was

Protein C - Biochemical and Medical Aspects

an inhibitor of coagulation and that it inactivated factor V (10). Subsequent reports indicated that activated protein C was more active against factor Va than factor V and that the enzyme was most active in the presence of phospholipids (11). Activated protein C has also been observed to inactivate factor VIII and factor VIIIa (12, 13, 14). Factor VIIIa appeared to be more sensitive to activated protein C than was factor VIII (14). In the presence of factor Xa, factor Va is protected from inactivation (10), suggesting that the prothrombinase complex is not inactivated by activated protein C. Not only does this anticoagulant enzyme exist in bovine plasma but it has also been shown to be present and active in human plasma (15). From studies on the effect of plasma components on the inactivation of factor Va by activated protein C, I have observed that an additional protein, protein S, is important for both the inactivation of factor Va and the anticoagulant activity of activated protein C (16).

II. Identification of protein S as a cofactor

The inactivation of factor Va by activated protein C proceeds optimally in the presence of calcium and phospholipids (10). If plasma is added, the rate of inactivation is observed to increase (16). The activity that stimulates the inactivation is referred to as the activated protein C cofactor activity. When plasma was absorbed with barium citrate the cofactor activity disappeared from the plasma. Chromatography and subsequent assay of the proteins that eluted from the barium citrate revealed that the cofactor activity eluted slightly ahead of the prothrombin peak. This elution position coincided with that of a protein that had been previously described, protein S (16). It was originally described by DiScipio and Davie who isolated this protein from human and bovine plasma (17). As they observed no function that could be ascribed to this protein, they named it protein S after the city of its discovery, Seattle. An alternative method for the purification of bovine protein S had also been described by Stenflo and Jonsson (18) and for the human protein by Dahlbäck (19). In addition to its effect on factor Va inactivation, protein S has also been recently observed to stimulate the inactivation of factor VIIIa in the presence of calcium and phospholipids (unpublished observations).

In order to determine if protein S was involved in the anticoagulant activity of activated protein C, a protein S deficient plasma was prepared. When activated protein C was added to this plasma, very little anticoagulant activity was observed (16). However, when protein S was restored to the plasma, the anticoagulant activity was also restored. Since the cofactor activity of protein S stimulates both factor Va inactivation and the anticoagulant activity of activated protein C, this is strong evidence supporting the conclusion that the mechanism of anticoagulation of activated protein C is by the proteolytic inactivation of factor Va (16).

III. Physical properties of protein S

Protein S is a vitamin K-dependent protein that has been purified from both human and bovine plasma (17, 18, 19). It contains a single polypeptide chain. When it is electrophoresed in polyacrylamide gel systems that contain the detergent sodium dodecyl sulfate, some heterogeneity can be observed. This heterogeneity exists in both the native and reduced forms of the protein. Some of the other physical properties of the protein are described in Table 1.

Table 1. Physical properties of protein S

Molecular weight:	
- SDS polyacrylamide gel electrophoresis	64,000
- Sedimentation equilibrium	60,000
Sedimentation coefficient	3.7
Stokes radius	43
A f/fo	1.5
Partial specific volume	0.729
Isoelectric point	5.4-5.9

One of the striking features of the protein is that it is the most basic of the vitamin K-dependent proteins. Protein S, like all of the vitamin K-dependent proteins, binds calcium ions. The binding of calcium ions to protein S leads to a quenching of the intrinsic fluorescence of the protein (20). Calcium will quench approximately 30 % of the intrinsic fluorescence at saturation. In addition to calcium, other divalent ions including barium, magnesium and manganese cause quenching. Protein S binds to phospholipid vesicles that contain acidic phospholipids. Among the vitamin K-dependent proteins, it has the highest affinity for acidic phospholipids (21).

In plasma, protein S appears to associate with the complement component C4b-binding protein. In human plasma, approximately 40 % of the protein S is found associated with the binding protein (22, 23, 24). It appears that slightly more protein S is associated with C4-bp in serum than in plasma. The functional significance of the interaction between protein S and C4-bp is unclear at this time (23).

IV. Functional properties of protein S

Protein S is the first of the vitamin K-dependent proteins to be described that does not appear to be a zymogen of a serine protease. There is no evidence that it has proteolytic activity either in the anticoagulant activity of activated protein C or in the stimulation of factor Va inactivation by activated protein C. Protein S has been isolated from a factor Va inactivation mixture in which activity was observed. Protein S did not have activity in the absence of activated protein C. Since protein S does not appear to behave as a protease, stimulation must be by an alternative mechanism. Titration of a factor Va inactivation mixture with protein S (with activated protein C, calcium and phospholipid held constant) indicated that the effect of protein S was saturable and that the stoichiometry of the interaction was approximately one. The apparent dissociation constant for the interaction between protein S and activated protein C was in the order of 1 nM suggesting that these two proteins have a high affinity for one another (25). In the absence of phospholipid, protein S had no effect upon the rate of inactivation of factor Va. Also, in the absence of lipid, there was no effect by protein S on the sedimentation behaviour of activated protein C, suggesting that the two proteins do not form a tight complex in the absence of phospholipid vesicles (20). Direct binding studies indicated that protein S increases the affinity of activated protein C for lipid vesicles by a factor of ten (25). Additional evidence that activated protein C (20) and protein S form a complex comes from the observations that either protein C or diisopropylfluorophosphate-labelled activated protein C (DIP-activated protein C) (25) can inhibit the effect of protein S on factor Va inactivation while having no effect on the rate of inactivation by activated protein C alone. This suggests that either protein C or DIP-activated protein C can form complexes with protein S, rendering it inactive (25).

Inactivation of factor Va by activated protein C and protein S have a complex dependence on the concentration of phospholipid as well as the types of lipids used. When the rate of inactivation is titrated with phospholipid, a biphasic effect on the rate is observed. At low concentrations, phospholipids stimulate the rate of inactivation. At high concentrations the rate is inhibited. Protein S appears to be most effective in stimulating inactivation at low lipid concentrations. The composition of the lipids also influences the rate of factor Va inactivation. When factor Va is inactivated with activated protein C (with calcium and phospholipids), the half maximal rate of inactivation is optimal with lipids composed of approximately 50 % phosphatidylserine, 50 % phosphatidylcholine. In the presence of protein S the half maximal rate shifts to around 15 % phosphatidylserine and 85 % phosphatidylcholine. By lowering the phosphatidylserine requirement it appears that protein S may be able to aid in the formation of the factor Va inactivation complex on membranes that contain low levels of phosphatidylserine. Phospholipid vesicles that contain only phosphatidylcholine had no effect upon the rate of factor Va inactivation by protein S and activated protein C.

V. Inactivation of protein S by thrombin

Thrombin is able to convert protein S from a single-chain protein to a two-chain protein (19, 26, 27). The two chains remain associated through disulfide bonds. Conversion of protein S into this form results in the loss of cofactor activity. Inactivation by thrombin can be blocked by the addition of calcium and phospholipids. Thrombin-cleaved protein S (two-chain protein S) has a number of altered properties in addition to the change in its primary structure. First, it has an altered interaction with calcium ions. Calcium ions can quench the intrinsic fluorescence of protein S approximately 30 %. However, there is virtually no effect of calcium on the intrinsic fluorescence of two-chain protein S (27). Calcium binding also appears to be reduced but not to the same degree as has been observed on the intrinsic fluorescence, suggesting that some of the calcium binding sites are intact in two-chain protein S. In addition, these results suggest that the relationship between calcium binding and the conformation change that causes quenching becomes uncoupled in the two-chain protein S. The interaction with lipid is also altered in two-chain protein S (26, 27). Native protein S has a high affinity for phospholipid, while two-chain protein S interacts very poorly. The loss of cofactor activity and lipid binding in two-chain protein S supports the hypothesis that protein S is involved in increasing the amount of activated protein C bound to phospholipid vesicles.

VI. Species specificity of protein S

When Marciniak reported that the treatment of a prothrombin preparation with thrombin could elicit an inhibitor of coagulation, she also observed that this inhibitor was extremely species-specific (28). That is, the bovine factor was only an anticoagulant in bovine plasma. There are several potential reasons why activated protein C might be species specific. One possibility is that there are species differences in factor Va that render it a poor substrate for activated protein C. A second possibility is that the species specificity exists in the interaction between activated protein C and protein S. To test this hypothesis I compared the anticoagulant activity of bovine activated protein C in human, rabbit, dog and bovine plasma. As had been reported previously, the bovine enzyme was a poor inhibitor of clotting in both rabbit and human plasma. It was an intermediate inhibitor in dog plasma and a good inhibitor in bovine plasma. When bovine protein S was added to each of the plasmas, the bovine enzyme became a potent inhibitor of human, rabbit and dog plasma. This result suggested that the reason for the species specificity resided in the protein S molecule. To confirm the result, bovine, human and rabbit plasma were assayed for activated protein C cofactor activity for factor Va inactivation. Only bovine plasma was observed to contain cofactor activity. These results indicate that the reason that activated protein C exhibits species specificity in its anticoagulant activity is that there is species specificity in the interaction between protein S and activated protein C (29).

This finding can be used to develop a simple one-stage clotting assay for protein S. Activated protein C from the species of protein S that one wishes to be measured and a plasma sample from a different species are required. The anticoagulant activity of activated protein C in that plasma sample will be directly proportional to the concentration of protein S added to the sample (20). An assay of this sort is useful in the purification of protein S, as one can ascertain by bioassay the quality of the material obtained. This is especially important since two-chain protein S will copurify with the single-chain protein. Without a bioassay, a protein S preparation that is mostly inactive might be obtained.

VII. Future trends

Since the discovery that protein S has cofactor activity for activated protein C, there has been an enhanced interest in the biology of protein C and proteins that are associated with its function. Though the biochemistry of protein S has been under scrutiny for several years, little is known about the physiological importance of either protein C or protein S as to their roles as regulators of thrombosis and hemostasis. The goals of future research should be to tie together clinical and biochemical findings in order to develop a picture of the physiology of this system and its role in the maintenance of blood fluidity.

Acknowledgement: This work was supported by grants from the National Institutes of Health (HL 26069) and the American Heart Association with funds provided in part by the Indiana Affiliate of the American Heart Association. The author is an Established Investigator of the American Heart Association.

References

1. Mammen, E.F., Thomas, W.R., Seegers, W.H.: Thromb. Diath. Haemorrh. 5, 218-249 (1960).

2. Seegers, W.H., Marlar, R.A., Walz, D.A.: Thromb. Res. 13, 233-243 (1978).

3. Seegers, W.H., McCoy, L.E., Groben, H.D., Sakuragawa, N., Agrawal, B.B.L.: Thromb. Res. 1, 443-460 (1972).

4. Seegers, W.H., Novoa, E., Henry, R.L., Hassouna, H.I., Thromb. Res. 8, 543-552 (1976).

5. Zolton, R.P., Seegers, W.H.: Thromb. Res. 3, 23-33 (1983).

6. Murano, G., Seegers, W.H., Zolton, R.: Thromb. Diath. Haemorrh. Suppl. 57, 305-314 (1974).

7. Stenflo, J.: J. Biol. Chem. 251, 355-363 (1976).

8. Esmon, C.T., Stenflo, J., Suttie, J.W., Jackson, C.M.: J. Biol. Chem. 251, 3052-3056 (1976).

9. Kisiel, W., Ericsson, L.H., Davie, E.W.: Biochemistry 15, 4893-4900 (1976).

10. Kisiel, W., Canfield, W.M., Ericsson, L.H., Davie, E.W.: Biochemistry 16, 5824-5831 (1977).

11. Walker, F.J., Sexton, P.W., Esmon, C.T.: Biochim. Biophys. Acta 571, 333-342 (1979).

12. Fay, P.J., Chavin, S.I., Schroeder, D., Young, F.E., Marder, V.J.: Proc. Natl. Acad. Sci. USA 79, 7200-7204 (1982).

13. Vehar, G.A., Davie, E.W.: Biochmistry 19, 401-410 (1980).

14. Marlar, R.A., Kleiss, A.J., Griffin, J.G.: Blood 59, 1067-1072 (1982).

15. Suzuki, K., Stenflo, J., Dahlbäck, B., Teodorsson, B.: J. Biol. Chem. 258, 1914-1920 (1983).

16. Walker, F.J.: J. Biol. Chem. 255, 5521-5524 (1980).

17 DiScipio, R.G., Davie, E.W.: Biochemistry 18, 899-904 (1979).

18. Stenflo, J., Jonsson, M.: FEBS Letters 101, 377-381 (1979).

19. Dahlbäck, B.: Biochem. J. 209, 837-846 (1983).

20. Walker, F.J.: Sem. Thromb. Haemostas. 10, 131-138 (1984).

21. Nelsestuen, G.L., Kisiel, W., DiScipio, R.G.: Biochemistry 17, 2134-2138 (1978).

22. Dahlbäck, B., Stenflo, J.: Proc. Natl. Acad. Sci. USA 78, 2512-2516 (1981).

23. Dahlbäck, B.: Biochem. J. 209, 847-856 (1983).

24. Dahlbäck, B., Smith, C.A., Müller-Eberhard, H.J.: Proc. Natl. Acad. Sci. USA 80, 3461-3465 (1983).

25. Walker, F.J.: J. Biol. Chem. 256, 11128-11131 (1981).

26. Suzuki, K., Nishioka, J., Hashimoto, S.: J. Biochem. 94, 699-705 (1983).

27. Walker, F.J.: J. Biol. Chem. 259, 10335-10339 (1984).

28. Marciniak, E.: Science 170, 452-453 (1970).

29. Walker, F.J.: Thromb. Res. 22, 321-327 (1981).

DOES HUMAN PROTEIN C EXERT PROFIBRINOLYTIC ACTIVITY?

Mario Colucci and Désiré Collen
Center for Thrombosis and Vascular Research, Department of Medical Research, University of Leuven, Belgium

The involvement of protein C (PC) in fibrinolysis was suggested more than 10 years ago by Seegers and coworkers (1). They showed that the addition of bovine autoprothrombin II-A (Auto II-A) (which is identical to activated PC (PCa)) to bovine or human blood, induced blood clot lysis. Auto II-A was devoid of plasminogen activator activity but was capable of reducing the activity of the inhibitors of fibrinolysis present in plasma (2). Infusion of Auto II-A into dogs also led to enhanced fibrinolysis as indicated by a shortening of the euglobulin lysis time (2).

Some years later, following the observation that the newly discovered protein C was identical to Auto II-A, additional investigations on the role of this protein in fibrinolysis were carried out. Taylor and coworkers reported that addition of bovine PCa to human blood produced a slight acceleration of the spontaneous lysis of blood clots (3). However, when PCa was added to a blood sample supplemented with low doses of plasminogen activator (PA), a very marked potentiation of the PA-induced clot lysis was observed, which could not be explained by an additive effect of the two enzymes (4). The profibrinolytic activity of PCa was not expressed in a plasma milieu. However, if plasma was enriched with adherent mononuclear cells, the PCa-induced acceleration of clot lysis was almost completely restored (3).

Protein C - Biochemical and Medical Aspects

Bovine PC has also been demonstrated to promote fibrinolysis in vivo. Comp and Esmon observed that injection of bovine PCa in dogs resulted in an enhanced rate of blood clot lysis which correlated with an increase of lysine adsorbable PA(s) in plasma (5). The authors concluded that PCa potentiates fibrinolysis by inducing the release of PA from endothelial cells. They also demonstrated that induction of PA release occurs via the generation of a second messenger in blood, most probably as the result of a proteolytic cleavage of an inactive precursor by PCa.

Comp et al. reported that infusion of low doses of thrombin in dogs resulted in the activation of protein C without apparent initiation of blood coagulation or platelet activation (6). Activation of endogenous PC by thrombin infusion was also associated with an increase of PA in the circulation. The finding that no PA response to thrombin could be observed in dogs anticoagulated with dicumarol suggested that PA release was caused by the endogenous PCa.

The data described above suggest that PC promotes fibrinolysis by at least two different mechanisms as schematically represented in Fig. 1. From the in vitro results it can be inferred that PCa enhances clot lysis either by potentiating the plasminogen activator-plasminogen system, thus leading to increased generation of plasmin, or by reducing the activity of the inhibitors of fibrinolysis such as coagulation factor XIII and α_2-antiplasmin. Monocytes, and probably other plasma proteins, play an essential role in this phenomenon. The in vivo experiments, on the other hand, indicate that PCa is capable of inducing the release of PA into the circulation. PCa would not be a PA release factor itself but generate a second messenger in blood, which in turn stimulates the secretion of PA from endothelial cells. Blood cells would participate in this phenomenon by enhancing the generation, or potentiating the activity, of this PA releasing factor.

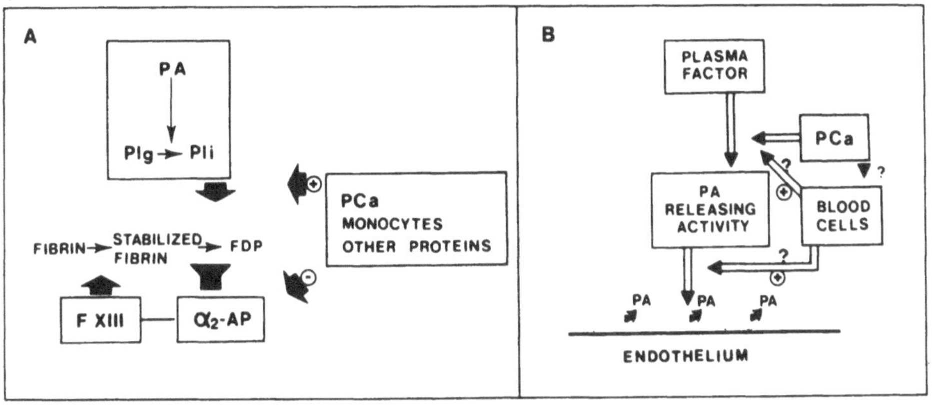

Fig. 1. Schematic representation of the proposed mechanisms of action of PCa. A. Possible interactions of PCa with the components of the fibrinolytic system; PA: plasminogen activator; PCa: activated protein C; Plg: plasminogen; Pli: plasmin; FDP: fibrin degradation products; F XIII: coagulation factor XIII; α_2-AP: α_2-antiplasmin. +: potentiation; -: inhibition. B. Induction of PA release by PCa.

Most of the present evidence on the profibrinolytic properties of PC is derived from experiments carried out with bovine PC. Earlier studies (7) have suggested that species variability may greatly influence the expression of the biological activity of PC. Therefore we have investigated the capacity of human PC to promote fibrinolysis in vitro and in vivo.

Profibrinolytic activity of human PC in vitro

Human PC was purified from factor IX concentrates as previously described (8) with slight modifications (9), and activated with human thrombin (10). The profibrinolytic capacity of human PCa was studied in an in vitro system consisting of radiolabelled crosslinked clots immersed in human plasma. Clots were prepared from whole human blood, platelet-rich plasma and platelet-poor plasma.

Addition of PCa (3 to 6 µg/ml) to the plasma surrounding the clots did not produce any significant modification of the rate of clot lysis as compared to control samples containing solvent or unactivated PC (Fig. 2). Furthermore PCa was unable to potentiate the rate of lysis induced by human tissue-type PA (t-PA) (2.5 to 5 IU/ml plasma). Similar negative results were obtained when PCa was incorporated into the clot.

These findings are in agreement with those obtained in rabbits with an experimental jugular vein thrombosis (unpublished). In these experiments infusion of human or bovine PCa (250 µg/kg body weight, over 4 hours) in rabbits with the experimental thrombus did not induce thrombolysis nor exert a synergistic effect on the thrombolysis induced by human t-PA (12,000 IU/kg body weight).

Profibrinolytic activity of human PC in vivo

Because of the species specificity of PC it is important, for the investigation of the in vivo effect on fibrinolysis, to select an animal species in which the biological properties of human PC are well expressed. We found the squirrel monkey (Samiri sciureus) to be a suitable model since a clear blood anticoagulation was induced in this primate by human PCa. Injection of human PCa (3 µg/ml blood), but not of unactivated PC, in anesthetized monkeys produced a prolongation of the activated partial thromboplastin time (APTT) without modification of the thrombin time (TT) (Fig. 3). Furthermore PCa did not alter the platelet number or the plasma levels of coagulation factor V and fibrinogen. Even a more pronounced anticoagulant effect was observed when a similar amount of human PCa was added to monkey plasma in vitro. Injection of PCa was, however, not associated with an increase of the fibrinolytic activity of blood (Fig. 3) as measured by the fibrin plate method (11).

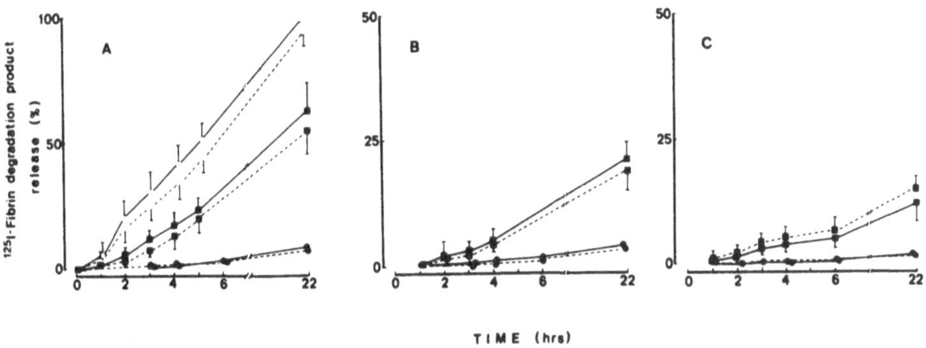

Fig. 2. Effect of human PCa on the lysis of ^{125}I-labelled clots in the presence and in the absence of human t-PA. Aliquots of citrated blood, platelet-rich plasma or platelet-poor plasma, containing ^{125}I-labelled fibrinogen, were clotted by addition of $CaCl_2$ (25 mM final concentration) and human thrombin (10 NIH U/ml). The mixtures were then aspirated in silicon tubes and allowed to clot for 30 min at 37°C. Pieces of 1 cm length were cut yielding labelled clots with a volume of about 130 µl. After washing in 0.15 M NaCl (30 min at room temperature with 3-4 changes) clots were immersed in 1.5 ml of plasma. Following the addition of test material or buffer (50 µl) the extent of lysis was calculated at the indicated intervals from the amount of radioactivity released into the plasma. Lysis of clots prepared from (A) whole blood, (B) platelet-rich plasma, and (C) platelet-poor plasma. (———), lysis in the presence of PCa (6 µg/ml); (- - -), lysis in the absence of PCa. (●), no addition of t-PA; (■), 2.5 IU t-PA/ml; (□), 5 IU t-PA/ml. Human t-PA was purified as described elsewhere (18).

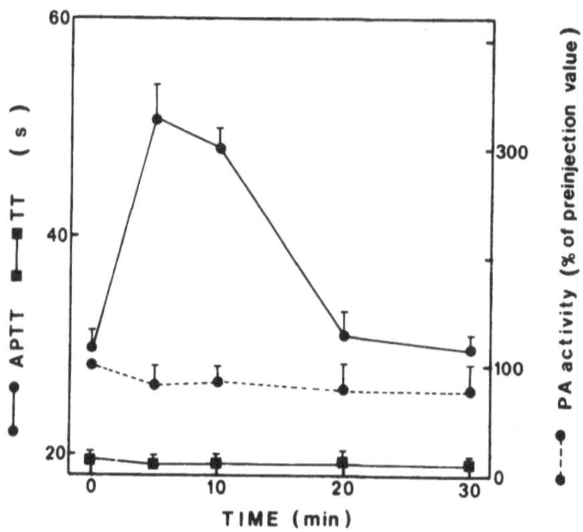

Fig. 3. Effect of human PCa injection (3 µg/ml blood) on plasma coagulation and PA activity. Squirrel monkeys (0.5-0.7 kg) were anesthetized by intramuscular injection of ketamine hydrochloride (25 mg/kg) and atropine (0.06 mg/kg) followed by intravenous injection of pentobarbital (30 mg/kg). Injections were given via a superficial foreleg vein and blood samples were taken by puncture from a femoral vein. Blood samples were anticoagulated with 4 % trisodium citrate (1/9 of the blood volume).

Previous investigations have shown that the generation of fibrinolytic activity by PCa is a complex phenomenon which involves cellular and humoral components. It is possible that human PC does not interact properly with the components involved in the fibrinolytic response of the squirrel monkey, which are different from the components involved in the anticoagulant response. We therefore studied the evolution of the blood fibrinolytic activity in monkeys in response to activation of endogenous PC.

Attempts to induce a selective activation of PC by infusion of low doses of thrombin (0.5 - 2 NIH U/kg/min over 20 or 60 min) were unsuccessful. This was evident from the lack of a selective prolongation of the APTT in the presence of an unchanged TT. Specific activation of PC could, however, be obtained by infusion of thrombin-thrombomodulin (T-TM) complex (10). It has been shown that thrombin, when bound to thrombomodulin, loses its coagulant and platelet activating activities (12, 13) and becomes a potent activator of PC (14). In our experiments the infusion of T-TM (2.9 µg/kg/min over 30 min), consisting of human thrombin and thrombomodulin purified from rabbit or monkey lungs (15), produced a marked prolongation of the APTT without change in the TT, plasma factor V levels, fibrinogen levels or platelet count (Fig. 4).

Blood anticoagulation was caused by activated PC as documented by the following observations. When plasma collected at the end of the T-TM infusion was added to normal monkey plasma, a prolongation of the APTT but not of the TT was observed, indicating that the anticoagulant effect produced by

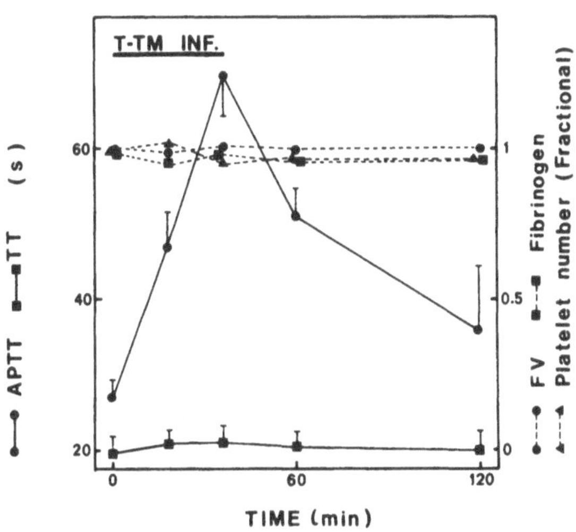

Fig. 4. Effect of T-TM infusion (2.9 µg/kg/min) on selected hemostatic parameters. The experimental procedure was as reported in the legend of Fig. 3

T-TM was not due to a deficiency of one or more of the clotting factors but to the generation of an anticoagulant activity which, like PCa, interfered with thrombin formation but not with thrombin activity. This anticoagulant activity slowly disappeared when plasma was incubated at 37°C (Fig. 5) which is compatible with the presence of a slow-reacting inhibitor of PCa in plasma (16).

The anticoagulant activity generated by T-TM infusion could be absorbed on barium citrate and recovered by elution with EDTA (10) (Fig. 6). The isolated anticoagulant was totally inhibited by rabbit IgG directed against human PC. In addi-

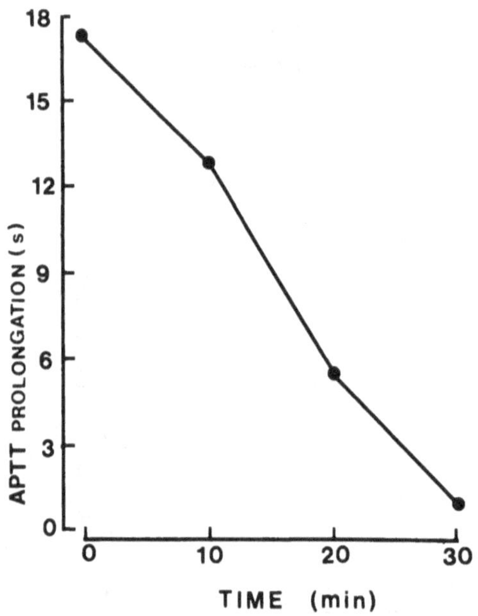

Fig. 5. Disappearance of the anticoagulant activity generated by T-TM infusion upon incubation of plasma (drawn at the end of the infusion) at 37°C. Twenty µl of the test plasma were added to 100 µl of normal monkey plasma (preactivated with kaolin for 5 min at 37°C) exactly 30 s before determination of the APTT.

tion it exhibited amidolytic activity towards the synthetic substrate S-2266 and was not inhibited by antithrombin III. The extent of PC activation in vivo by T-TM was estimated by comparing the anticoagulant and amidolytic activities recovered in a barium citrate eluate of postinfusion plasma to the anticoagulant and amidolytic activities generated in a similar preinfusion sample following complete activation of PC by thrombin in vitro. Both assays revealed that the peak levels of PCa activity following T-TM infusion corresponded to 20 - 30 % activation of circulating PC.

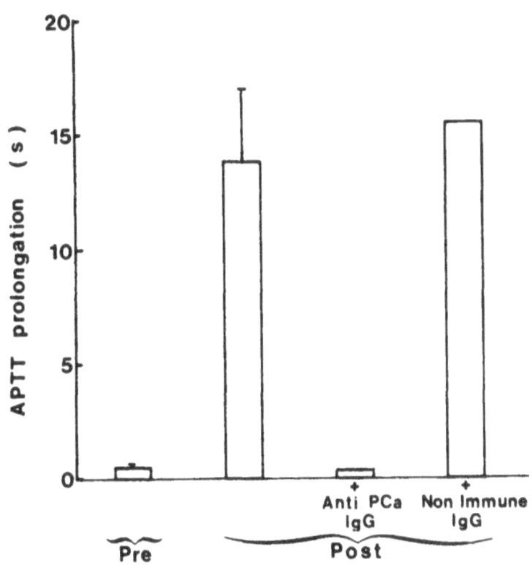

Fig. 6. Anticoagulant activity in barium citrate eluates of plasma samples collected before (pre) and at the end (post) of T-TM infusion. Quenching experiments were carried out by incubating the barium citrate plasma eluates with anti-human PC rabbit IgG or non-immune rabbit IgG (0.5 mg/ml final concentration) for 2 h at 37°C. The anticoagulant activity of the test samples was measured as outlined in the legend of Fig. 5.

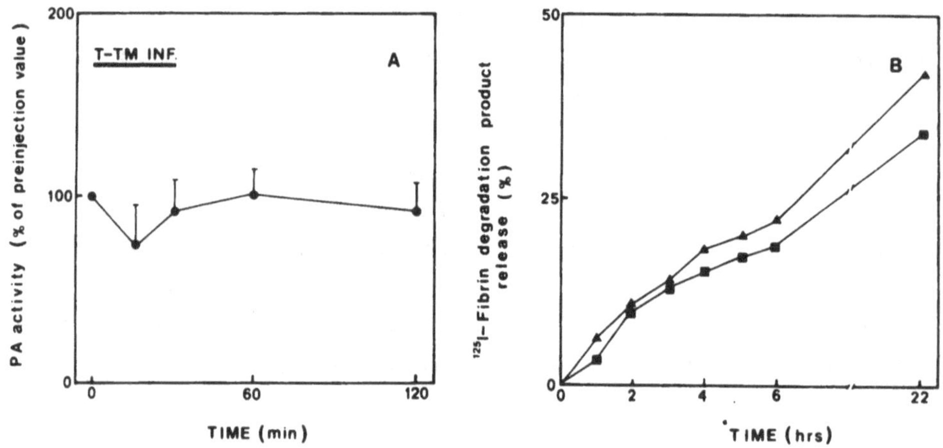

Fig. 7. Fibrinolytic response to T-TM infusion. A. PA activity of the euglobulin fraction of plasma samples collected at the indicated intervals following the start of the infusion. B. ^{125}I-fibrin-split-product release from clots prepared from whole blood samples collected before (▲) and immediately after (■) the end of T-TM infusion.

Notwithstanding the fact that T-TM induced a selective and marked activation of PC in the squirrel monkey, no change in the PA activity of the euglobulin fraction nor in the rate of lysis of whole blood clots was observed (Fig. 7).

Comp and Esmon (5) showed that, in dogs, the release of PA in response to PCa occurs via a second messenger present in blood. It is possible that the squirrel monkey lacks this intermediate messenger. The following experiments were carried out to assess whether PCa is capable of inducing generation of PA releasing activity in human blood. Human blood was incubated with PCa (6 µg/ml) for 10 min at 37°C. Five ml of plasma obtained from this latter sample (corresponding to 20 % of the total plasma volume of the squirrel monkey) were then infused into the animal and blood samples were taken at intervals for PA assay. No fibrinolytic response (PA activity ranging from 80 to 105 % of the preinfusion value) could be observed for one hour following the plasma infusion. Such evidence suggests, but does not conclusively establish that no second messengers for PCa-induced PA release are generated in human blood.

The lack of fibrinolytic response to PCa cannot be due to a poor responsiveness of the squirrel monkey to PA releasing agents. We have previously observed (17) that injection of known stimulators of PA release, such as arginine-vasopressin or its synthetic analog 1-desamino 8-D-arginine-vasopressin (DDAVP), induced a marked fibrinolytic response in the squirrel monkey. The increase of PA activity in blood following DDAVP stimulation is due to the release of t-PA. More than 80 % of the PA activity of the euglobulin fraction of plasma obtained from animals injected with DDAVP was, indeed, quenched by rabbit Ig directed against human t-PA, whereas less than 5 % was inhibited by a similar concentration of anti-urokinase rabbit Ig or non-immune Ig (10).

In physiological conditions T-TM complexes are formed on the surface of the endothelium and do not circulate free in the blood (14). To exclude the theoretical possibility that circulating T-TM complexes, as encountered in our experimental conditions, were interfering with the release of PA, we measured the fibrinolytic response to DDAVP in monkeys that had received an infusion of T-TM. Injection of DDAVP (10 µg/kg) to animals immediately after the end of T-TM infusion induced a clear PA response which was comparable to that observed in untreated animals following DDAVP stimulation (10).

Conclusions

Under our experimental conditions we did not observe a profibrinolytic activity of human PCa both in an in vitro system and in squirrel monkeys in vivo. Previous studies had shown that injection of bovine PCa in dogs enhances the fibrinolytic activity of blood. We therefore conclude that human PC does not exert a similar profibrinolytic activity as bovine PC.

References

1. Seegers, W.H., McCoy, L.E., Groben, H.D., Sakuragawa, N., Agrawal, B.B.L.: Thromb. Res. 1, 443-460 (1972).
2. Zolton, R.P., Seegers, W.H.: Thromb. Res. 3, 23-33 (1973).
3. Taylor, F.B., Esmon, C.T., Carrol, R.C., Locart, M.L.: Thromb. Haemostas. 46, 61 (1981).
4. Taylor, F.B., Carrol, R.C.Jr., Gerrard, J., Esmon, C.T., Radcliffe, R.D.: Fed. Proc. 40, 2092-2098 (1981).
5. Comp, P.C., Esmon, C.T.: J. Clin. Invest. 68, 1221-1228 (1981).
6. Comp. P.C., Jacocks, R.M., Ferrel, G.L., Esmon, C.T.: J. Clin. Invest. 70, 127-134 (1982).
7. Marciniak, E.: J. Lab. Clin. Med. 79, 924-934 (1972).
8. Suzuki, K., Stenflo, J., Dahlbäck, B., Teodorsson, B.: J. Biol. Chem. 258, 1914-1918 (1983).
9. Sala, N., Owen, W.G., Collen, D.: Blood 63, 671-675 (1984).
10. Colucci, M., Stassen, J.M., Collen, D.: J. Clin. Invest. 74, 200-204 (1984).
11. Astrup, T., Müllertz, S.: Arch. Biochem. Biophys. 40, 346-351 (1952).

12. Esmon, C.T., Esmon, N.L., Harris, K.W.: J. Biol. Chem. 257, 7944-7947 (1982).

13. Esmon, N.L., Carrol, R.C., Esmon, C.T.: Thromb. Haemostas. 50, 81 (1983).

14. Esmon, C.T., Owen, W.G.: Proc. Natl. Acad. Sci. USA 78, 2249-2252 (1981).

15. Esmon, N.L., Owen, W.G., Esmon, C.T.: J. Biol. Chem. 257, 859-864 (1982).

16. Suzuki, K., Nishioka, J., Hashimoto, S.: J. Biol. Chem. 258, 163-168 (1983).

17. Colucci, M., Stassen, J.M., Salwa, J., Collen, D.: Br. J. Haematol. (in press).

18. Rijken, D.C., Collen, D.: J. Biol. Chem. 256, 7035-7041 (1981).

BIOCHEMISTRY AND PHYSIOLOGY OF PROTEIN C INHIBITOR

Koji Suzuki
Department of Laboratory Medicine, Mie University School of Medicine, Tsu City, Mie 514, Japan

Introduction

The report in 1980 of Marlar and Griffin (1) which indicated the deficiency of the inhibitor of activated protein C (APC), designated protein C inhibitor (PCI), in plasma of patients with congenital combined deficiency of coagulation factor V and factor VIII have stimulated the discussion of the role of protein C and its inhibitor in regulating blood coagulation. That report also encouraged us to isolate the PCI from human plasma.
Two years later, PCI was purified from human plasma and characterized both chemically and immunologically (2, 3).
The present article describes on the biochemistry and physiology of human PCI, and then the synthesis and secretion of the inhibitor by hepatoma-derived cell line (4).

Biochemical properties of human PCI

PCI, purified from human plasma by the method as described in the previous report (2), is a single polypeptide chain with M_r 57,000 on SDS-polyacrylamide gel electrophoresis under both unreduced and reduced conditions.

Table 1. Biochemical properties of protein C inhibitor

Molecular weight (single polypeptide)	
SDS-PAGE	57,000
gel filtration	55,000
$E_{1cm}^{1\%}$ at 280 nm	14.1
Isoelectric points (microheterogeneity)	7.4 - 8.5
Electrophoretic migration	$\alpha_1 - \alpha_2$-globulin
Inhibition spectrum	
(+)	activated protein C thrombin factor X a trypsin chymotrypsin
(-)	plasmin urokinase
Affinity for heparin	

Tab. 1 shows the biochemical properties of PCI. PCI differed immunologically from the already established protease inhibitors in plasma, e.g. α_1-antitrypsin, α_1-antichymotrypsin, α_2-macroglobulin, α_2-plasmin inhibitor, C_1-esterase inhibitor, antithrombin III and heparin cofactor II. The amino acid composition of PCI was also different from those of other plasma protease inhibitors (2). PCI seemed to be a basic amino acid-rich protein. Fig. 1 shows the amino-terminal sequences of PCI (5) and other heparin-binding proteins, APC-binding protein (6), antithrombin III (7) and heparin cofactor II (8). There was no structural homology among the terminal sequences of these proteins. PCI showed uniquely basic isoelectric points between pH 7.4 and 8.5 with microheterogeneity (2).

PCI has rather high affinity for heparin. Fig. 2 shows an affinity chromatography of PCI and APC-binding protein on the column of heparin-agarose.

APC-binding protein and PCI were separately eluted out from the column with the buffer containing 0.2 - 0.3 M NaCl and 0.35 - 0.45 M NaCl, respectively, on NaCl gradient elution.

	1 5 10 15
Protein C inhibitor	His-Arg-His-His-Pro-Arg-Glu-Met-Lys-Lys-Arg-Val-Glu-Asp-Leu-
APC-binding protein	Gly-Arg-Thr-Cys-Pro-Lys-Pro-Asp-Asp-Leu-Pro-Phe-Ser-Thr-Val-
Heparin Cofactor II	Gly-Ser-Lys-Gly-Pro-Leu-Asp-Gln-Leu-Glu-Lys-Gly-Gly-Glu-Thr-
Antithrombin III	His-Gly-Ser-Pro-Val-Asp-Ile-Cys-Thr-Ala-Lys-Pro-Arg-Asp-Ile-

Fig. 1. Amino-terminal sequences of PCI, APC-binding protein, heparin cofactor II and antithrombin III.

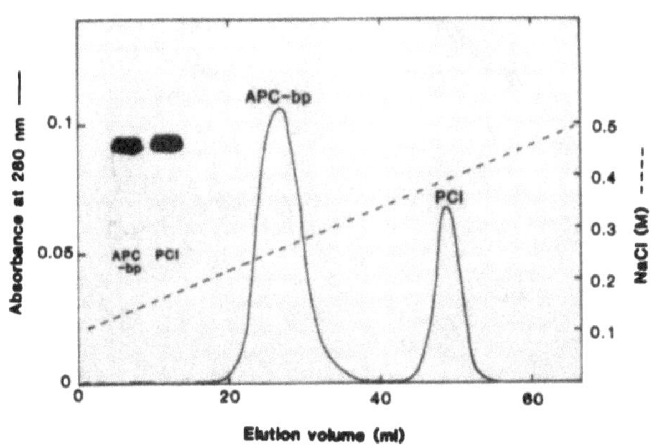

Fig. 2. Affinity chromatography of PCI and APC-binding protein on heparin-agarose column. PCI (200 µg) and APC-binding protein (500 µg, given from Dr. W. Kisiel) were mixed in 0.05 M Tris-HCl, 0.1 M NaCl, pH 7.5, and then applied to the heparin-agarose column equilibrated with the same buffer noted before. Proteins were eluted from the column with NaCl gradient (0.1 - 0.5 M). PCI and APC-binding protein treated beforehand with 5 % 2-mercaptoethanol were analyzed on SDS-polyacrylamide gel electrophoresis.

Effect of heparin on the action of PCI

The activity of PCI which inhibits the amidolytic activity of APC was stimulated by heparin (3). Fig. 3 shows the effect of varying concentrations of heparin on the second-order rate constants of the reaction between APC and PCI. The rate constants reached maximum at a concentration of 5 to 10 units heparin per ml. It is noticeable that the rate of the APC inhibition by PCI was accelerated about 30-fold in the presence of optimal concentration of heparin. The inhibition rates of PCI against thrombin, factor Xa, trypsin and chymotrypsin were also increased by the presence of heparin.

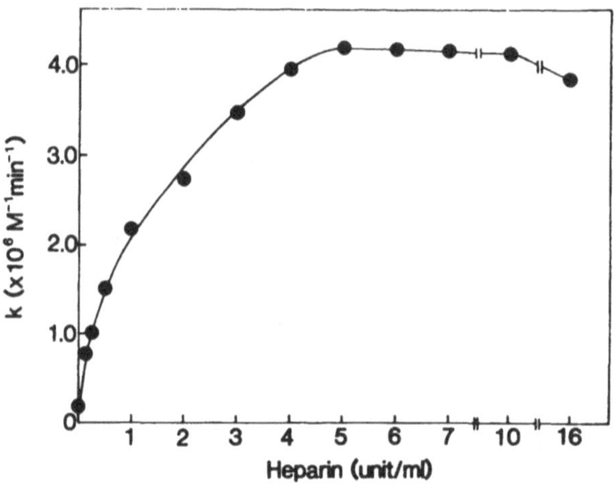

Fig. 3. Second-order rate constants for the inhibition of APC by PCI in the presence of heparin. APC (0.54 µg) and PCI (6.0 µg) in 0.05 M Tris-HCl, 0.1 M NaCl, pH 8.0 (Tris-buffered saline) containing 0.1 % bovine serum albumin were incubated at 37°C for 1 min in the presence of various concentrations of heparin. 2.5 ml of 100 µM Boc-Leu-Ser-Thr-Arg-MCA (peptide-MCA substrate) in Tris-buffered saline containing 2 mM $CaCl_2$ was then added to the incubation mixture, and the residual activity of the enzyme was measured by initial rate method.

Heparin accelerated the inhibition of APC by plasma PCI as well. In the absence of heparin the half-life of APC in plasma was about 30 min, whereas heparin shortened it to about 1 min.

Other polysaccharides except for heparin were examined whether they effect the reaction between APC and PCI. However, none of chondroitin sulfate A and C, dermatan sulfate, heparan sulfate and hyaluronic acid showed any noticeable effects on the reaction. Furthermore, the several synthetic sulfated polysaccharides, namely dextran sulfate, were tested. As shown in Fig. 4, some dextran sulfates with approximately 18 % sulfur content accelerated the reaction between the enzyme and the inhibitor. These results suggest

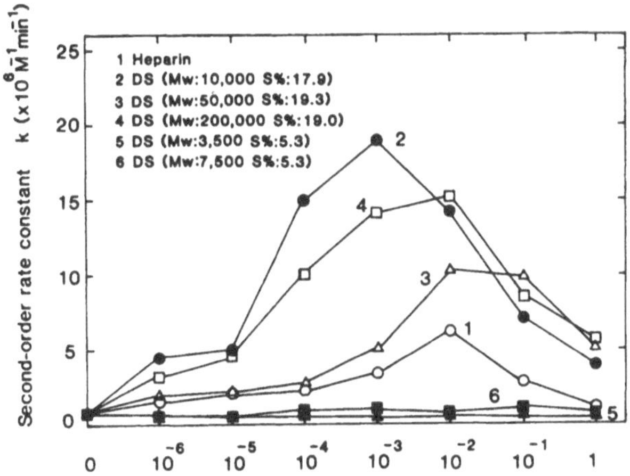

Fig. 4. Effect of dextran sulfates and heparin on the reaction between APC and PCI. APC (0.54 µg) and PCI (6.0 µg) in Tris-buffered saline containing 0.1 % bovine serum albumin were incubated at 37°C for 1 min in the presence of various concentrations of dextran sulfate (DS) or heparin. 2.5 ml of 100 µM peptide-MCA substrate in Tris-buffered saline containing 2 mM $CaCl_2$ was then added to the incubation mixture, and the residual activity of enzyme was measured. The rate of the reaction is indicated as second-order rate constants.

that the sulfated polysaccharides including heparin bind to PCI and induce the conformational change of the inhibitor to accelerate the inhibition toward the enzyme.

Murine monoclonal antibodies against human PCI, which had been prepared according to the method of Köhler and Milstein (9), were tested about the effect on the biological properties of PCI. As shown in Fig. 5, one of these antibodies inhibited the stimulation effect of heparin on PCI action. However, this antibody did not inhibit the stimulation effect of the dextran sulfate. These facts suggest the presence of the heparin-specific binding site on PCI molecule, and that the dextran sulfates do not necessarily bind to the heparin-specific binding site to stimulate the PCI activity.

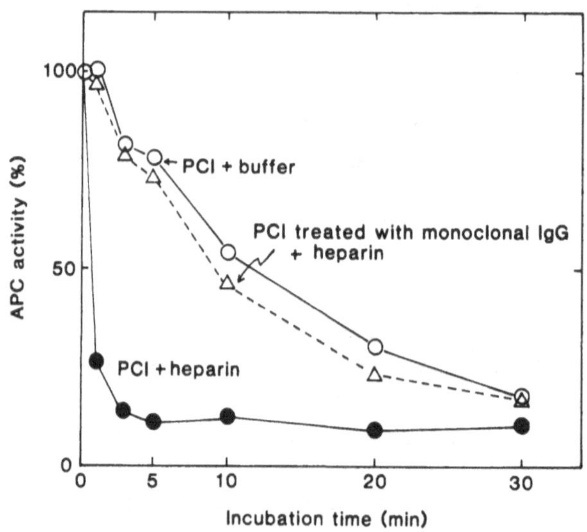

Fig. 5. Effect of monoclonal antibody on the inhibition of APC by PCI in the presence of heparin. 20 µl of APC (0.75 µg) was incubated with 20 µl of PCI (2.0 µg), which had been treated with the monoclonal antibody (10 µg) or Tris-buffered saline containing 0.1 % bovine serum albumin overnight, in the presence or absence of heparin (5 units/ml). At intervals, the residual activity of APC was measured by using peptide-MCA substrate.

Inhibition mechanisms of APC by PCI

The mechanism of inhibition of APC by PCI was studied by using SDS-polyacrylamide gel electrophoresis. As shown in Fig. 6, the band of the complex, formed between APC and PCI, with M_r 102,000 in unreduced sample and the band with M_r 88,000 in reduced sample were observed after incubation of APC and PCI. In addition to these bands, a band with M_r 54,000 appeared.

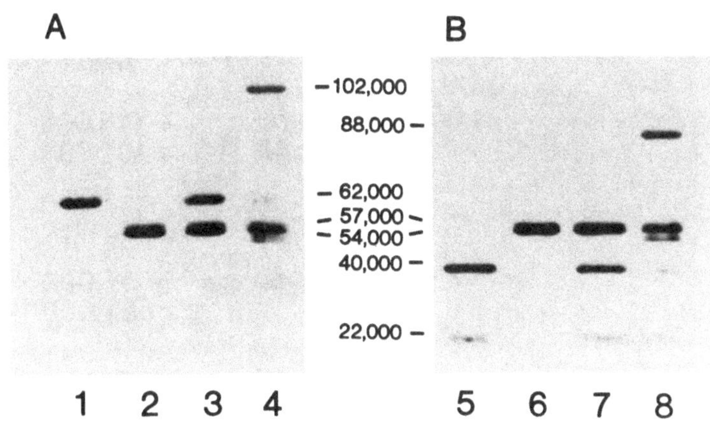

Fig. 6. SDS-polyacrylamide slab gel electrophoresis of the reaction mixtures of APC and PCI, unreduced (A) and reduced (B). 10 µg of APC was incubated with 15 µg of PCI in Tris-buffered saline at 37°C for 30 min. The samples, unreduced or reduced with 5 % 2-mercaptoethanol, in 1 % SDS and 18 % sucrose, were subjected to SDS-polyacrylamide slab gel electrophoresis. 1 and 5, APC; 2 and 6, PCI; 3 and 7, mixture of APC and PCI after 30 min incubation. M_r of each band was estimated by comparison with the standard proteins (Pharmacia Fine Chem., Sweden).

The M_r change of PCI during the APC inhibition was then studied by using ^{125}I-labelled PCI. The labelled PCI, which had been incubated with varying amounts of APC, were subjected to SDS-polyacrylamide slab gel electrophoresis and then to autoradiography (Fig. 7). The band with M_r 118,000 would be a dimerized inhibitor, because of its disappearance after treating with 2-mercaptoethanol. According to the increase of APC added, the intact PCI band (M_r 57,000) was converted to a little smaller M_r band with M_r 54,000 concomitantly with the formation of a complex with an apparent M_r 102,000 in unreduced samples. In reduced samples, M_r of the complex changed to 88,000, but the band with M_r 54,000 still remained at the same position as observed

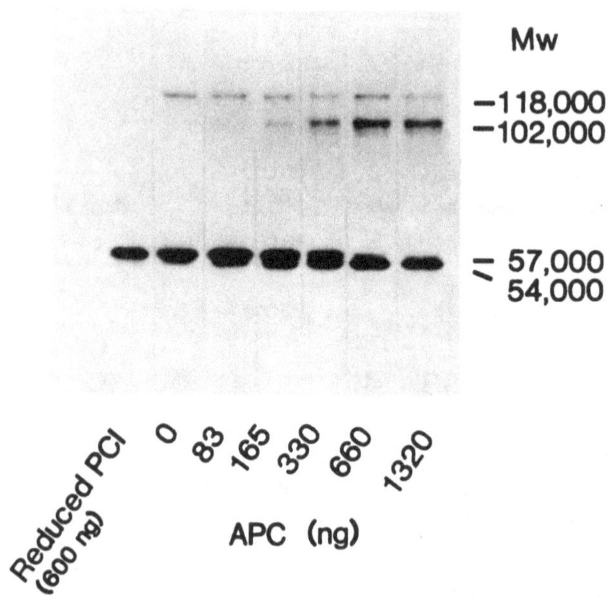

Fig. 7. Autoradiography of SDS-polyacrylamide slab gel containing samples of the ^{125}I-PCI and APC. 20 µl of ^{125}I-PCI (0.6 µg) was incubated with 20 µl of various amounts of APC (noted in figure) in Tris-buffered saline containing 0.1 % bovine serum albumin and heparin (5 units/ml) at 37°C for 15 min. The samples were subjected to SDS-polyacrylamide slab gel electrophoresis and then to autoradiography. The sample 1 was reduced, and the samples from 2 to 7 were unreduced.

in unreduced samples. These complex formation and the modification of PCI were not observed when unactivated protein C or DIP-APC was used.

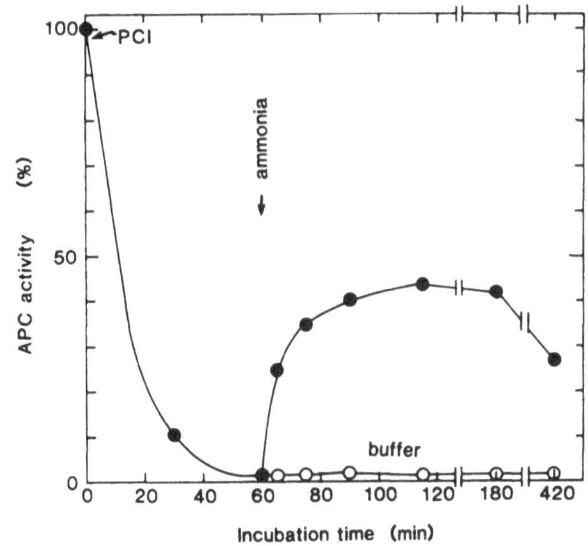

Fig. 8. Effect of ammonia on the complexes formed between APC and PCI. 200 µl of APC (10 µg) in Tris-buffered saline containing 0.1 % bovine serum albumin was incubated with 50 µl of PCI (10 µg) in the same buffer at 37°C. After 60 min, 250 µl of 2 M ammonia was added to the reaction mixture and incubated at 37°C. At intervals, 25 µl aliquots were taken from the mixture and the activity of APC was measured by using peptide-MCA substrate. The original activity of APC was taken as 100 %.

The effects of nucleophilic agents, e.g. ammonia and hydroxylamine, on the complex between APC and PCI were examined in order to elucidate the nature of the bond between the enzyme and the inhibitor. After the APC activity was completely inhibited by an almost equimolar PCI, the incubation mixture was treated with 1 M ammonia or hydroxylamine. Fig. 8 shows the effect of ammonia. Approximately 40 % of the APC activity was released from the complex by the ammonia treatment. Almost the same recovery of the APC activity was shown by treating the complex with hydroxylamine.

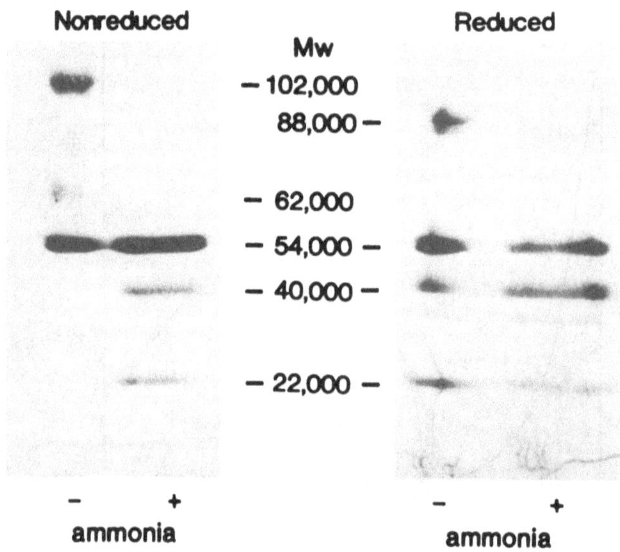

Fig. 9. SDS-polyacrylamide slab gel electrophoresis of the reaction mixture of APC and PCI before and after treating with ammonia. 16 µg of APC was incubated with 15 µg of PCI in Tris-buffered saline containing heparin (5 units/ml) at 37°C for 30 min. The reaction mixture, unreduced (lanes 1 and 2) and reduced with 5 % 2-mercaptoethanol (lanes 3 and 4), was then treated with 1 M ammonia in 1 % SDS and 18 % sucrose for 30 min. The samples were subjected to SDS-polyacrylamide slab gel electrophoresis. Lanes 1 and 3, without ammonia treatment; lanes 2 and 4, with ammonia treatment.

The dissociation of the complex with hydroxylamine was then investigated by treating with 1 M ammonia by using SDS-polyacrylamide slab gel electrophoresis (Fig. 9). Each complex band in unreduced and reduced samples was found to dissociate into two subunits of the free enzyme and a free modified inhibitor with M_r 54,000. These results indicate that PCI is proteolytically modified during the complex formation. The nature of bond between the modified PCI and APC is very similar to that between the modified antithrombin III and thrombin, factor Xa or factor IXa as reported so far (10, 11). It is probably an acyl-bond.

Simple model for the mechanism of inhibition of APC by PCI is illustrated in Fig. 10. In the course of the APC inhibition, PCI forms the complex with APC and at the same time PCI is proteolytically modified by the enzyme. Then a part of the modified inhibitor appears to dissociate from the complex, though the mechanism of the dissociation is still unknown.

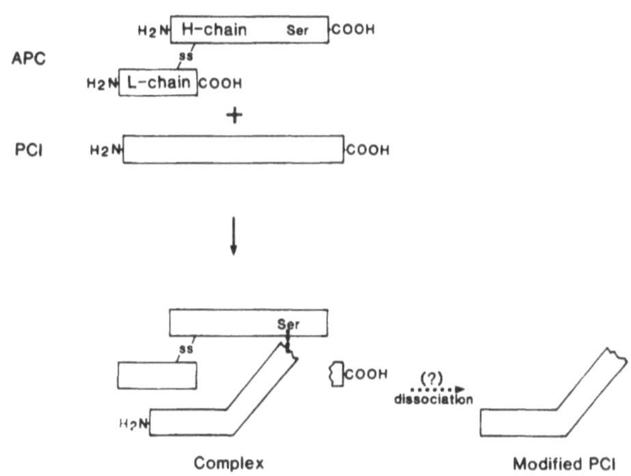

Fig. 10. Proposed model for mechanism of inhibition of APC by PCI

Physiology of PCI

The immunological experiments *in vitro* indicate that PCI is the only plasma inhibitor of APC, because the PCI-immunodepleted plasma, which had been prepared by using anti-

PCI-IgG coupled agarose, did not show any detectable inhibition on APC (2, 3). On the other hand, PCI inhibited thrombin, factor Xa as well as APC in the test tube when purified enzyme reacted with the inhibitor (3). Therefore it was examined whether PCI reacted with thrombin and/or factor Xa during blood coagulation by immunoassay (EIA) and functional APC-inhibition assay. Both the concentration of immunoreactive PCI and the functional activity of PCI in serum were identical to those of PCI in plasma. This fact suggests that PCI does not react with any activated coagulation factors during blood coagulation *in vitro* (12).

As described previously (5, 13), the clinical significance of PCI in pathogenesis of congenital combined deficiency of factor V and factor VIII might become less important. However, since APC appears to be a potent anticoagulant and profibrinolytic agent, PCI may play an important role in the regulation of coagulation and fibrinolysis. Therefore, the concentration of PCI was measured as well as protein C in plasma of normal individuals and patients with liver disease or intravascular coagulation or patients who were treated with warfarin. Liver disease was evident by positive bilirubin (over 2 mg/dl) and also positive GOT and GPT (over 100 Karmen units). Intravascular coagulation was evidenced by positive FDP (over 10 µg/ml serum), and the samples of the group with intravascular coagulation were selected from the patients with normal liver function.

The patients who had been treated with warfarin were chosen from the patients showing the Thrombotest activity below 20 % of normals. Fig. 11 shows the level of protein C and PCI antigen in plasma of normals and the patients. The level of PCI decreased in plasma of patients with liver disease regardless of the presence or absence of FDP. The concentration of plasma PCI in patients with intravascular coagulation distributed widely. Some decreased but some did not. The level of PCI in these patients appeared to change in parallel with that of protein C in the same patients. On the

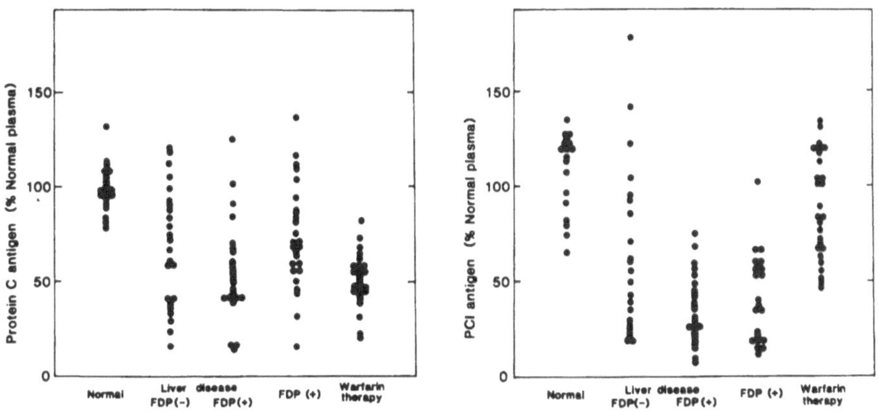

Fig. 11. Level of protein C and PCI antigen in plasma of normal individuals and patients with liver disease, intravascular coagulation or warfarin therapy.

other hand, the level of PCI in patients with warfarin therapy did not reduce essentially, although the level of immunoreactive protein C reduced significantly. These results suggest that the level of PCI does not always decrease depending on that of protein C, but presumably depends on the activation of protein C accompanying intravascular coagulation. The correlation between PCI and protein C in plasma was studied in normal individuals and patients with intravascular coagulation and normal liver function. The correlation coefficient between both proteins was 0.64 ($p < 0.001$, $n = 47$). Because the liver function in these groups appeared to be normal, this significantly high correlation suggests that PCI is consumed only when protein C is activated during blood coagulation *in vivo*. Similar correlations in the same plasma samples were shown between PCI and antithrombin III ($r = 0.73$, $p < 0.001$) and between PCI and factor V ($r = 0.57$, $p < 0.001$). However, the correlation between PCI and factor VIII was low ($r = 0.34$, $p < 0.01$).

Synthesis and secretion of PCI in hepatoma cells

The fact that the level of plasma PCI was reduced in patients with liver cirrhosis suggests that PCI is produced in the liver. Then the study on the synthesis and secretion of PCI by the human cultured liver cells was performed (4). The cell line Hep G2, which has been established to produce and secrete many liver-specific plasma proteins including the components in blood coagulation and fibrinolysis (14, 15, 16), was used for this purpose. The PCI produced and secreted into the culture medium was determined by immunoassay (RIA) and also APC inhibition assay. Fig. 12 shows the production and secretion of PCI from Hep G2. Accompanying the increase of cell number, the concentration of immunoreactive PCI in the medium increased progressively during 13 days of the culture. The PCI produced was able to inhibit the amidase activity of APC as well, and its specific activity was identical to that purified from plasma.

The lysate of Hep G2 cells did not contain a significant amount of PCI. This result suggests a small intracellular pool and the rapid secretion of PCI as it is synthesized. The *de novo* synthesis of PCI by Hep G2 was further confirmed by the incorporation of ^{35}S-labelled methionine into the PCI protein. The data obtained from the examination using hepatoma cells show that the liver is a site of PCI production.

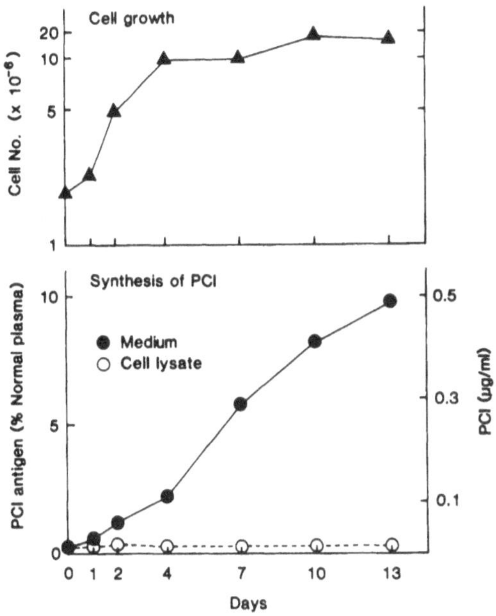

Fig. 12. Time course of cell growth and accumulation of PCI in the culture medium. 1.8 x 10^6 Hep G2 cells were plated in 75 mm^2 culture flask with 15 ml of the new medium.
A: cell number; B: concentration of PCI antigen in the medium or in cell lysate.

Acknowledgement: The author wishes to thank Professor Hashimoto S. and Mr. Nishioka J. for their generous support throughout the study. The original work described in this article was supported in part by Grants from the Scientific Research Fund of the Ministry of Education, Science and Culture of Japan, Japan Heart Foundation and the Research Foundation for Cancer and Cardiovascular Disease of Japan.

References

1. Marlar, R.A., Griffin, J.H.: J. Clin. Invest. 66, 1186-1189 (1980).

2. Suzuki, K., Nishioka, J., Hashimoto, S.: J. Biol. Chem. 258, 163-168 (1983).

3. Suzuki, K., Nishioka, J., Kusumoto, H., Hashimoto, S.: J. Biochem. 95, 187 - 195 (1984).

4. Morita, F., Saito, H., Suzuki, K., Hashimoto, S.: In contribution.

5. Suzuki, K.: Semin. Thromb. Haemostas. 10, 154-161 (1984).

6. Canfield, D.M., Kisiel, W.: J. Clin. Lab. 70, 1260-1272 (1982).

7. Tollefsen, D.M.: Personal communication.

8. Petersen, T.E., Dudek-Wojciechowska, G., Sottrup-Jensen, L., Magnusson, S.: Physiological Inhibitors of Blood Coagulation and Fibrinolysis, Elsevier, Amsterdam, p 43-54 (1979)

9. Köhler, G., Milstein, C.: Nature 256, 495-497 (1975).

10. Fish, W.W., Björk, I.: Eur. J. Biochem. 101, 31-38 (1979).

11. Björk, I., Jackson, C.M., Jörnall, H., Lavine, K.K., Nordling, K., Salsgiver, W.J.: J. Biol. Chem. 257, 2406-2411 (1982).

12. Suzuki, K., Nishioka, J., Yamada, Y., Hashimoto, S.: The XIIth World Congress of Anatomic and Clinical Pathology (abst.), Tokyo (1983) p 293.

13. Suzuki, K., Nishioka, J., Hashimoto, S., Kamiya, T., Saito, H.: Blood 62, 1266-1270 (1983).

14. Knowles, B.B., Howe, C.C., Aden, D.P.: Science 209, 497-499 (1980).

15. Saito, H., Goodnough, L.T., Knowles, B.B., Aden, D.P.: Proc. Natl. Acad. Sci. USA 79, 5684-5687 (1982).

16. Fairs, D.S., Plow, E.F.: J. Lab. Clin. Med. 101, 372-384 (1983).

ISOLATION OF PROTEIN C USING HIGH PERFORMANCE LIQUID CHROMATOGRAPHY (HPLC) - EXPERIENCES AND PROBLEMS

Chieko Miyashita, Peter Hellstern, Gerhard von Blohn, Hans Hammer, Ernst Wenzel

Abteilung für Klinische Hämostaseologie und Transfusionsmedizin, Universität des Saarlandes, D-6650 Homburg-Saar

Introduction

The purification of human protein C (PC) was first described by Kisiel in 1979. In the meantime additional methods of purification have been reported by Suzuki (1983) and by Comp (1984). The common starting point for all three methods is the adsorption of human citrated platelet poor plasma (cPPP) or ACD-PPP to barium chloride (Tab. 1).

Table 1. Principles of human PC purification methods described by other authors.

Procedure	Kisiel 1979	Suzuki 1983	Comp 1984
Material	cPPP	ACD-PPP	ACD-PPP
Adsorption	$BaCl_2$ 1 x	$BaCl_2$ 1 x	$BaCl_2$ 2 x
Precipitation Ammonium Sulf.	35 %, 65 %	-	30 %, 60 %
Chromatography	DEAE-Seph. A50 dextran sulf.agar.	DEAE-Sephacel heparin-Sephar.	QAE-Sephadex
Electrophor.	disc	-	-
Determination	aPTT/S-2238	Va inactiv.	ester. activity

Protein C - Biochemical and Medical Aspects

Precipitation with ammonium sulfate was performed by Kisiel and by Comp. Various chromatographic procedures were utilized. After chromatography (CHR), preparative disc electrophoresis was applied by Kisiel. The determination of PC in the purified fractions was conducted by measuring the prolongation of the activated partial thromboplastin time (aPTT), by S-2238, by factor V inactivation, or by measuring the esterase activity of PC.
Although the HPLC method is possibly faster and more efficient in separation, it has to be ascertained whether different surfaces with unknown characteristics as well as the higher pressure can be applied to PC purification.

Methods

A computerized LKB high performance system with pump (2150), controller (2152), and monitor (2151), equipped with DEAE TSK column was used. 2 ml of starting solution were chromatographed at a rate of 300 µl/min and detected at 280 nm. The flow was kept constant. The salt gradient was also controlled. The single steps of PC purification are shown in Tab. 2. Adsorption of PC was obtained with 1 M $BaCl_2$. The sediment was dissolved in the eluting buffer and directly chromatographed without precipitation with ammonium sulfate. Albumin was removed by filtering through a column of Blue Sepharose CL/6B. The albumin-free material was further chromatographed on DEAE TSK 545. We did not add protease inhibitors during the CHR procedure. We also did not use affinity CHR after ion exchange CHR. Obviously our buffer systems were different from those of other authors.

Table 2. Simplified description of human PC isolation using HPLC

Starting material: ACD-PPP
Adsorption: (1 M $BaCl_2$)
Elution 50 mM imidazole, benzamidine, EDTA; Blue Sepharose
Elution PBS pH 7.4
Dialysis
DEAE-TSK 545
Elution 50 mM imidazole pH 6.0 gradient 0 - 1 M NaCl
Determination of protein C: ELISA, S-2366

Results

Up to and including the Blue Sepharose step all steps of our modifications were identical. In the first modification a phosphate buffer (0.1 M, pH = 6.0), a salt gradient of 0 - 0.6 M NaCl, and the column DEAE TSK 545 were used.
The second modification consisted of the elution buffer Tris/HCl (20 mM, pH = 8.0), a salt gradient of 0 - 1 M NaCl, and the same column (DEAE TSK 545). Rechromatography was done on a DEAE TSK 5 PW using a piperazin buffer (20 mM, pH = 9.0) and a salt gradient of 0 - 1 m NaCl.
The third modification included DEAE TSK 545, imidazole/HCl buffer (50 mM, pH = 6.0) and a salt gradient of 0 - 1 M NaCl. The procedure was repeated two times.
As expected the first modification was not very successful because of disturbances of chromogenic substrate assays and of coagulation assays due to the phosphate buffer. In con-

trast to Suzuki, who could use another buffer system, the efficacy of separation was also unsatisfactory. The second modification demonstrated a satisfactory separation (Fig. 1). However, in 10 runs a significant decrease of PC activity was observed in contrast to the high PC concentration. In the third modification the alteration of the pH to 6.0 reduced the efficacy of protein separation whereas the PC activity increased. The elution diagram and protease activities using S-2366, S-2238, and S-2222 are shown in Fig. 2. Since S-2366 is not completely specific and may also be cleaved by thrombin, a decisive statement cannot be made about the purity of PC based on these findings. In order to attain more exact assessment, factors II, VII, and X were determined by clotting assays. The results are shown in Tab. 3. The contamination by these coagulation factors was high.

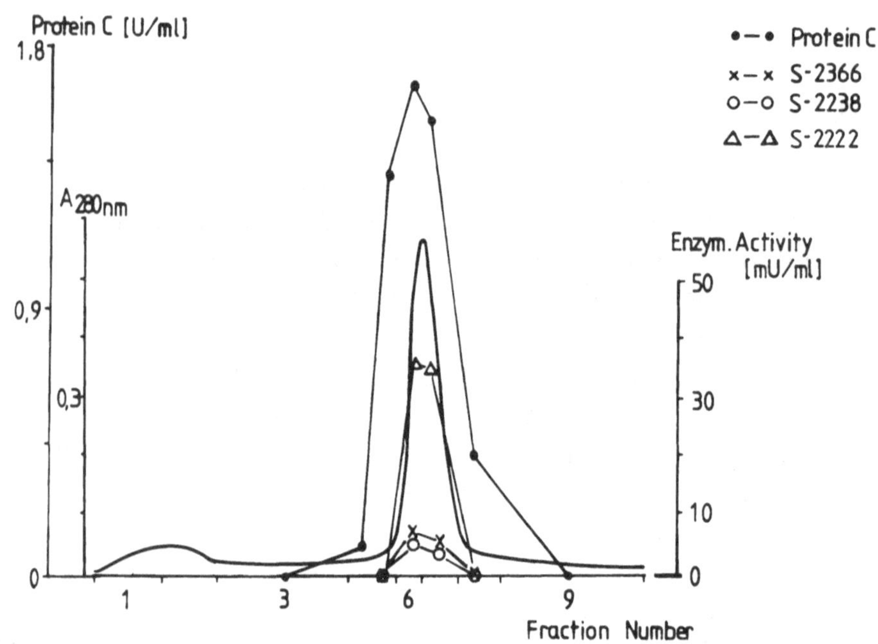

Fig. 1. Elution diagram, PC concentration, and proteolytic activities of the PC-rich fraction produced by the second modification of PC isolation using HPLC.

Table 3. Composition of the PC-rich fraction produced by the third modification of PC isolation using HPLC

	U/ml	
Protein C	0.14	ELISA
F II	0.03	one-stage clotting assay
F VII	0.01	
F IX	0.00	
F X	0.02	

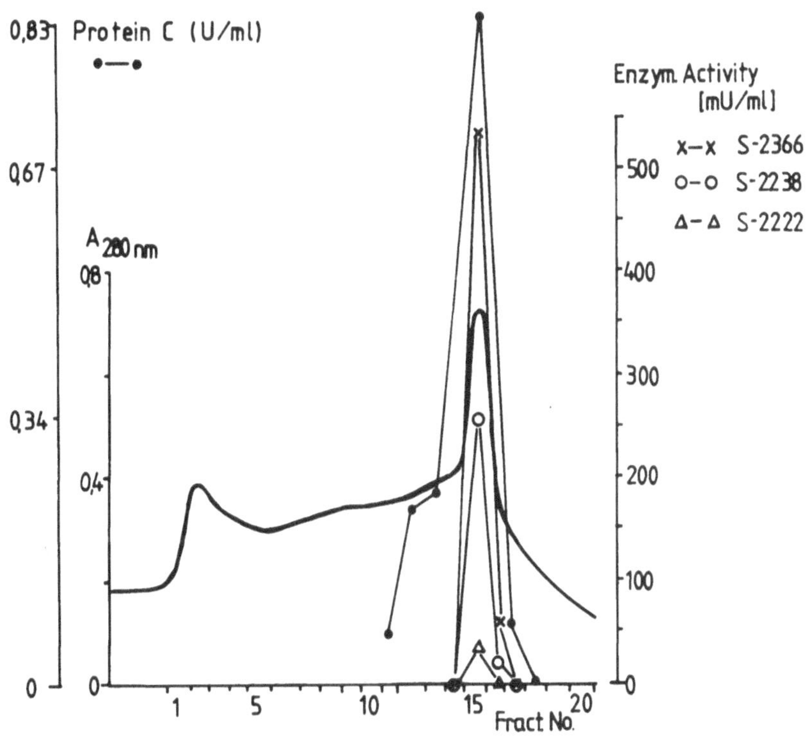

Fig. 2. Elution diagram, PC concentration, and proteolytic activities of the PC-rich fraction produced by the third modification of PC isolation using HPLC.

Conclusions

1. In comparison to conventional methods the following preparatory steps before, during, and after CHR can be summarized in three simple modifications:

   a) We did not precipitate with ammonium sulfate because we observed high PC losses.
   b) We separated the PC-rich fraction from albumin and perhaps parts of prothrombin before HPLC CHR. Without this step albumin contamination was found in all fractions.
   c) In comparison to other chromatographic methods we suspect that it is not necessary to add protease inhibitors when applying HPLC due to the short passage and rapid separation.

2. As expected the HPLC method is faster. Two to 3 days are necessary for isolation rather than 5 - 6 days with conventional methods. No more than 2 hours are required for one elution with HPLC. In our experience PC can be isolated using DEAE TSK 545, imidazole buffer (pH = 6.0), and a 0 - 1 M NaCl salt gradient. A greater puritiy is achieved by repeating the procedure two times. This involves a prolongation of only 4 hours without significant loss of PC.

3. In future studies we shall include affinity CHR to achieve improved separation and greater specific activities. This is necessary because Suzuki, using another method of PC activity determination, found nearly a tenfold greater specific activity. One possible cause of low PC activities could be an unsatisfactory relationship betweeen surface area and sample volume. This has to be investigated further.

References

1. Comp. P.C., Nixon, R.R., Esmon, C.T.: Blood 63, 15-21 (1984).

2. Kisiel, W.: J. Clin. Invest. 64, 761-769 (1979).

3. Suzuki, K., Stenflo, F., Dahlbäck, R., Teodorsson, B.: J. Biol. Chem. 258, 1914-1920 (1983).

DETERMINATION OF PROTEIN C BY AN IMMUNOENZYMATIC ASSAY

Jean Amiral
Lab. Diagnostica Stago, 95130 Franconville, France

Catherine Boyer, Chantal Rothschild, Martine Wolf
Lab. Central d'Hématologie, Hôpital de Kremlin Bicêtre
94270 Le Kremlin Bicêtre, France

Introduction

Protein C is an anticoagulant protein of major importance in regulating in vivo blood coagulation. In its activated form it inactivates factors Va and VIIIa (1, 2). Measurement of this factor has allowed diagnosis of congenital or acquired deficiencies. Low levels of protein C have been found to be associated with recurrent thrombosis (3, 4). Since it is now possible to purify protein C (5, 6), specific antibodies can be developed and used for immunological assays. In respect to the low concentration of protein C in human blood (3-5 µg/ml) very sensitive methods have to be considered. For this purpose, a rabbit serum to human protein C was developed into rabbits and used for immunological determination. Two methods were investigated and found useful for routine clinical applications: the Laurell method (Electro Immuno Diffusion), previously described by others for this application (4, 7, 8), and a new ELISA (Enzyme Linked Immuno Sorbent Assay) (9). The latter achieves good sensitivity and low detection thresholds.

Purification of human protein C

Human protein C was purified from 12 liters of citrated normal human plasma, kept in the frozen state until use. After

Protein C - Biochemical and Medical Aspects

thawing at 37°C, plasma was supplemented with protease inhibitors, to prevent activation of coagulation factors, then cooled at 4°C. The general procedure used for the purification was the technique described by Kisiel (6), with some minor modifications. Barium citrate was added under continuous stirring. The prothrombin complex was adsorbed on the insoluble barium citrate precipitate formed. This precipitate was collected by centrifugation and then extensively washed. Desorption of protein was achieved with ammonium sulfate. The eluate was salt fractionated with ammonium sulfate and the prothrombin complex obtained in the 40-70 % fraction. It was then desalted by gel filtration on a column of Sephadex G 25 (Pharmacia, Sweden), equilibrated with the initial buffer used in the next step. The prothrombin complex was chromatographed on DEAE Sephacel. Elution was performed with a sodium chloride salt gradient from 0.2 to 0.6 M in phosphate buffer (0.05 M) at pH 6.0. The first peak, collected following initiation of the gradient, was factor VII. Protein C eluted at about 0.3 M, whereas protein S, factor X, factor IX and prothrombin were eluted with higher salt concentrations. Further purification of protein C consisted in affinity chromatography on Dextran Sulfate Agarose (10) in a citrate buffer at pH 6.00. Elution was performed with a sodium chloride salt gradient. Protein C eluted at 0.15 M. If necessary, the two last steps were run again. Alternatively, while prothrombin and factor X were the main measurable contaminants, immunoadsorption was done on anti-factor X agarose column and anti-prothrombin agarose column. The final product was controlled and found to be without factor VII, II, IX or X activities in detectable antigenic amounts, whereas the protein C concentration was about 0.100 mg/ml, according to absorbance at 280 nm ($E_{1\%}^{1cm}$ = 14.5 (6)), or 25 U/ml as compared to a normal pooled plasma (1 U/ml). Studies of this product using SDS polyacrylamide gel electrophoresis, with or without reduction, showed the classical pattern: only one band corresponding to a Mr of 62,000 daltons for unreduced gels and two

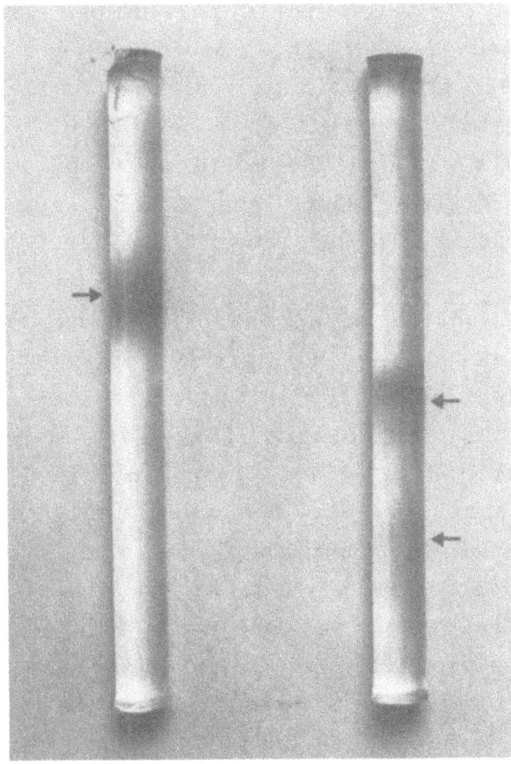

Fig. 1. SDS-PAGE of purified protein C. Unreduced gel is on the left and reduced gel on the right.

bands of Mr 41,000 and 21,000 respectively corresponding to heavy and light chain of protein C (Fig. 1).
Routinely, about 5 mg of protein C were obtained. This protein was used for immunization of rabbits with conventional methods (11).

Production of specific antibodies

Rabbits were brought to the hyperimmune state by repeated immunizations with pure protein C. This state was usually reached three months after starting immunization. Rabbits were then bled every two weeks for an one year period. All the sera collected over this period were pooled and controlled for reactivity with protein C. Usually an adsorption step was necessary to remove trace amounts of non-specific antibodies. This was done by solid phase adsorption using an agarose beads column, where the various fractions obtained during purification and devoided of protein C were covalently linked (12).

Specificity of antiserum was checked using double diffusion Ouchterlony's method where antiserum was opposed to normal human plasma and pure protein C, immuno-electrophoresis, and bidimensional electrophoresis (13, 14) with normal human plasma or protein C deficient plasma (obtained by solid phase immuno-adsorption). In all the cases only a precipitating line was obtained with normal plasma whereas no immuno-precipitation was observed with deficient plasma. Two different sources of deficient plasma were used; one was a kind gift of Dr. Bertina (Leiden, The Netherlands) and the other was prepared in the laboratory. In addition, this antiserum showed no inhibitory effect on factors II, VII, IX or X activities when incubated with normal human plasma.

This two dimensional electrophoresis was found to be the most sensitive way to control specificity as increasing amounts of antisera were included in the second migration step (up to 20 %). This resulted in a dramatic increase in sensitivity to detect contaminants (Fig. 2).

The process used to produce antibodies (long immunization period and pooling of all the sera collected over this hyperimmune period) resulted in antibodies with good characteristics: high avidity, high titer, good reactivity to the complete protein structure. This antiserum reacts with the normal protein C as well as with that with the hypo- or acarboxylated form present in patients on oral anticoagulant therapy. Both forms can be separated by electrophoresis in the presence of Ca^{++} (9, 15). The normal protein C binds Ca^{++} and has retarded migration. Fig. 3 shows a bidimensional electro-

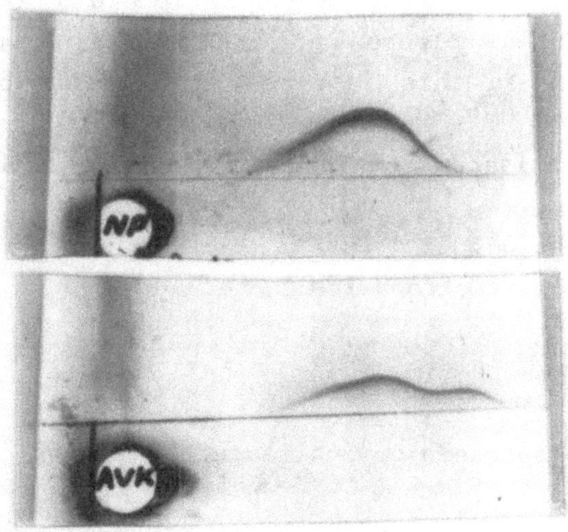

Fig. 2. Control of specificity of anti-protein C rabbit serum using bidimensional electrophoresis. Normal human plasma always produces only one precipitating line whereas deficient plasma shows no immunoprecipitation.

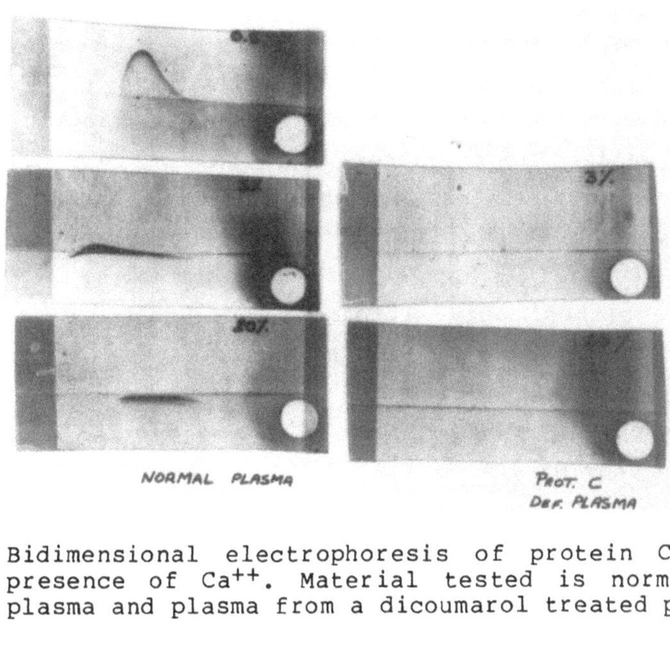

Fig. 3. Bidimensional electrophoresis of protein C in the presence of Ca^{++}. Material tested is normal human plasma and plasma from a dicoumarol treated patient.

phoresis of protein C, in presence of Ca^{++} in the first migration of normal plasma and a plasma from a dicoumarol treated patient. In this last case both forms of protein C are present.
This specific antiserum obtained was used for several immunological methods:

- Laurell's method,
- Bidimensional electrophoresis,
- ELISA method.

Determination of protein C with Laurell's method

Despite the low plasma concentration of protein C, Laurell's method (13) has been successfully used. Agarose A 37 (IBF, France) is used at 1 % in a tris glycine barbital buffer (16) containing 5 mM Na_2EDTA. This chelating agent is necessary in order to avoid differences in electrophoretic mobility due to interaction with Ca^{++} (especially when determining levels of protein C on dicoumarol treated patients). Anti-protein C rabbit serum (ASSERA Protein C) is used at a 0.3 % concentration. 15 µl of serial dilutions of calibration plasma (1/1 to 1/8) or tested plasmas are applied in 4 mm diameter wells. Electrophoresis is run for 16 to 20 hours at 1.5 to 2 mA/cm. Plates are washed, dried and stained with Coomassie brillant blue R 250. Usually normal plasma produces a rocket 20 to 25 mm height.
With this method, only protein C concentrations above 10 to 15 % can be accurately measured. However, this technique is suitable for small series and sufficient to detect protein C deficiencies. A ready-to-use kit, ASSERA-PLATE Protein C, allows routine determination.

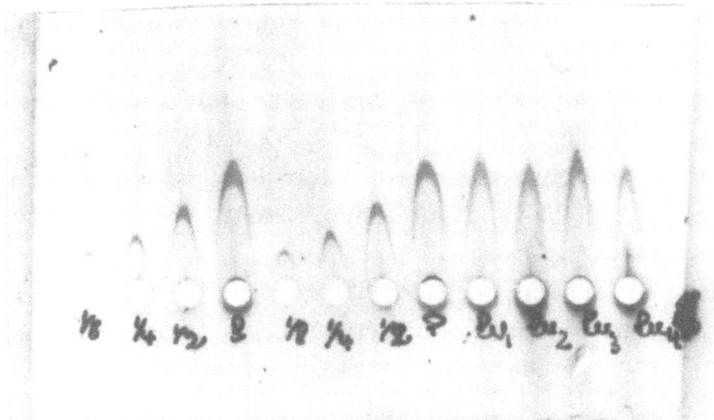

Fig. 4. Laurell's determination of protein C.

ELISA sandwich method for protein C

The ELISA sandwich method has been used to measure macromolecular antigens (with at least two different and spaced epitopes), with high sensitivity (17, 18). Pure antibodies must be purified from specific antiserum and used either for coating purposes or to prepare the enzyme antibody conjugate. To coat the plates, $(F(ab')_2$ fragments are preferred to whole immunoglobulins, as it avoids interference of rheumatoid factor (19). Anti-protein C antibodies are labelled with the peroxidase method reported by Nakane (20). Reagents are stabilized, standardized and lyophilized. Kits ELISA PROTEIN C (Boehringer Mannheim, Germany) and ASSERACHROM PROTEIN C are commercially available for this ELISA measurement of protein C.

The various steps involved in this determination are:

- Coating of a solid phase support (polystyrene ELISA plate) with $F(ab')_2$ fragments specific for protein C (overnight at room temperature), then washing.

- Addition of dilutions of calibration plasma or test plasma and incubation for two hours at room temperature, then washing.
- Addition of an excess of conjugate and incubation at room temperature for two hours, then washing.
- Addition of ortho phenylene diamine with hydrogen peroxide and stopping the colour development three minutes later with 3 M sulfuric acid.
- Reading of absorbance at 492 nm.

A dose-response curve is obtained on a large working range (from the 1/50 to the 1/5000 dilutions of normal plasma). It is convenient to draw it on log-log paper (Fig. 5).

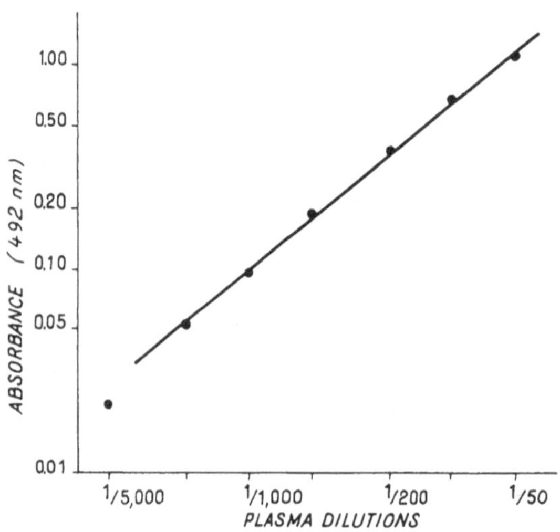

Fig. 5. Dose-response curve for calibration of measurement of protein C with ELISA method.

As the ELISA method is far more sensitive than all the gel precipitation methods (detection threshold in the ng/ml range as compared to thresholds in the 1 - 10^{-1} µg/ml range), sensitivity of the test must be checked again. For this purpose the two deficient plasmas previously described have been used, at 1:5, 1:20 and 1:50 dilutions. Even the 1:5 dilution gives an absorbance similar to the one obtained for buffer alone, and lower than the one corresponding to the 1:5000 dilution of normal plasma. So no interference of other plasma proteins is observed in this test with a high specificity for human protein C.

This ELISA method has some main advantages. It can be done within 5 hours. It has a low detection threshold (0.0002 U/ml of protein C in the test sample, i.e. 1 ng/ml). It is accurate (CV ranging from 3 to 7 % for intra-series and from 4 to 9 % for inter-series, depending on the protein C concentration). In addition high automation possibilities exist. So the method is quite adequate for large series.

Tab. 1 compares the performance characteristics of the ELISA method versus Laurell's.

Comparison between Laurell's and ELISA methods

Protein C was determined on various groups of plasmas with both methods. All samples were collected in 0.11 M citrate (1 vol. per 9 vol. whole blood) and plasmas obtained by centrifugation at 3000 rpm for 15 minutes, and kept frozen until use. Reference plasma consisted of a pool of 20 normal plasmas, stored frozen in small aliquots. The samples studied were plasmas from:

- 40 normal individuals (selected on the basis of normal hemostasis and absence of underlying disease).
- 14 congenital protein C deficiencies (kindly provided by Drs. R. Bertina, Leiden, The Netherlands, M. Samama, Paris, France, and J.P. Cazenave, Strasbourg, France).
- 40 dicoumarol treated patients.
- 15 patients suffering from biopsy proven liver cirrhosis.

Tab. 2 compares the levels of protein C antigen by ELISA and by Laurell's method and the correlation coefficients obtained. In all the groups a good correlation is observed (Fig. 6).

Table 1. Comparison of performance characteristics of ELISA and Laurell's methods for protein C determination.

	ELISA	LAURELL
Detection limit	0.0002 U/ml (0.02 %)	0.1 U/ml (10 %)
Accuracy: CV	3 - 9 %	5 - 10 %
Time required	5 h	24 h
Automation	High possibilities	No
Equipment	Reader (492 nm) Micro Plates Washer Multichannel Pipette	Electrophoresis Power Supply Electrophoresis Chamber

Table 2. Determination of protein C by ELISA and EID in normal and pathological plasmas.

Plasmas	N	ELISA	EID	r*
Normals	40	91.3 ± 17.1	90.8 ± 18.7	0.93
Congenital	13	41.2 ± 8.1	37.1 ± 9.3	0.89
deficiencies	1	9	13	-
Liver cirrhosis	15	18.9 ± 6.3	18.4 ± 4.6	0.97
Dicoumarol treated patients	40	44.1 ± 15.4	44.8 ± 16.2	0.92

* correlation coefficient: ELISA/Laurell

It is concluded that for the various groups there is no preferred method, and both give similar results for determination of protein C.

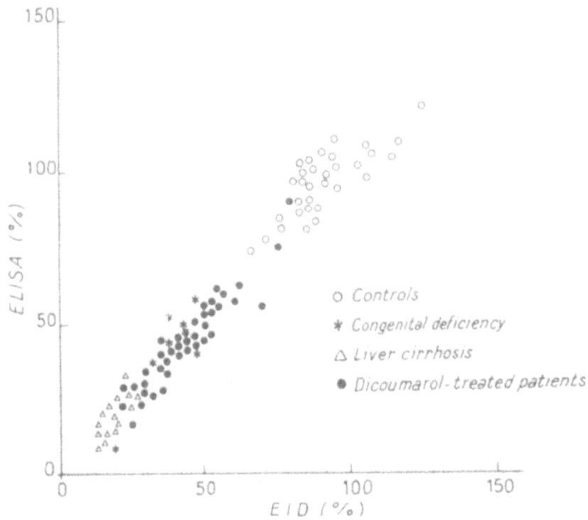

Fig. 6. Correlation between levels of protein C obtained by ELISA and bei EID.

Conclusion

Since there is evidence that the protein C system plays a major role in anticoagulation it has become necessary to have practical methods for clinical laboratory investigation. It would be interesting to have a global test investigating the whole system and additional determinations of each parameter. Effectively recurrent thrombosis may theoretically result from various defects: lack of classical inhibitors (AT III...), deficiencies in the protein C system (protein C, protein S, thrombomodulin, ...), deficiencies in fibrinolysis (plasminogen, tissue plasminogen activator, ...). Only some of these are investigated in current practice (AT III, plasminogen, fibrinolysis). Determination of protein C may give a response to some unexplained pathologies. Because of this immunological tests are of clinical importance. Recently, functional defects have been described when protein C activity has been measured by some authors (21, 22, 23, 24). Nevertheless, these assays are not fully standardized and, depending on the principle used, they measure more or less completely the protein C system (Bertina, in this volume). There are, therefore, numerous applications for ELISA and Laurell's method for protein C determination. Besides diagnosis of congenital deficiencies, acquired deficiencies of protein C can be investigated (7). In oral anticoagulant therapy, protein C (as factor VII), which has a short half-life, decreases more rapidly than other factors (II, IX, X) (9, 15). It induces a transitory thrombotic state in the oral anticoagulant therapy. In addition poor or no correlation exists between variations of the various vitamine K-dependent factors. Extreme cases can be found with high levels of procoagulant factors and low levels of protein C and vice versa. It can be thought that determination of protein C in oral anticoagulant therapy may be a helpful index concerning the patient's protection against thrombosis.

For the hemostasis laboratory, it is necessary to work with practical tests. As they are now commercially available, ELISA and Laurell's methods can be introduced for routine use. Choice will depend on the series done, the availability of specific equipment and the sensitivities needed. This new parameter, protein C, can be introduced in routine clinical laboratories, with reproducible and standardized methods.

References

1. Marlar, R.A., Kleiss, A.J., Griffin, J.H.: Ann. N. Y. Acad. Sci. 370, 303-310 (1981).
2. Nesheim, M.E., Canfield, W.M., Kisiel, W., Mann, K.G.: J. Biol. Chem. 257, 1443-1447 (1982).
3. Griffin, J.H., Evatt, B., Zimmerman, T.S., Kleiss, A.J.: J. Clin. Invest. 68, 1370-1373 (1981).
4. Bertina, R.M., Broekmans, A.W., Van der Linden, I.K., Mertens, K.: Thromb. Haemostas. 48, 1-5 (1982).
5. Stenflo, J.: J. Biol. Chem. 251, 355-363 (1976).

6. Kisiel, W., Canfield, W.M., Ericsson, L.H., Davie, E.W.: Biochemistry 16, 5824-5831 (1977).

7. Mannucci, P.M., Vigano, S.: Lancet, Aug. 28, 463-467 (1982).

8. Pabinger-Fasching, I., Bertina, R.M., Lechner, K., Niessner, H., Korninger, Ch.: Thromb. Haemostas. 50, 810-813 (1983).

9. Boyer, C., Rothschild, C., Wolf, M., Amiral, J., Meyer, D., Larrieu, M.J.: Thromb. Res. 36, 579 - 589 (1984).

10. Pepper, D.S., Prowse, C.: Thromb. Res. 11, 687-692 (1977).

11. Harboe, N., Ingild, A.: Scand. J. Imm. 2, Suppl. 1, 161-164 (1973).

12. March, S.C., Parikh, I., Cuatrecasas, P.: Anal. Biochem. 60, 149-152 (1974).

13. Laurell, C.B.: Anal. Biochem. 15, 45-52 (1966)

14. Ganrot, P.O.: Scand. J. Clin. Lab. Invest. 29, Suppl. 124, 39-41 (1972).

15. Vigano, S., Mannucci, P.M., Solinas, S., Bottasso, B., Mariani, G.: Brit. J. Haemat. 57, 213-220 (1984).

16. Weeke, B.: Scand. J. Imm. 2, Suppl. 1, 15-35 (1973).

17. Engvall, E., Perlmann, P.: J. Immunol. 109, 129-135 (1972).

18. Avrameas, S., Guilbert, B.: Biochimie 54, 837-842 (1972).

19. Hagenaars, A.M., Knipers, A.W., Nagel, J.: Dev. Clin. Biochem. 1, 209-215 (1979).

20. Nakane, P.K., Kawaoi, A.: J. Histochem. Cytochem. 22, 1084-1091 (1974).

21. Francis, R.B., Patch, M.J.: Thromb. Res. 32, 605-613 (1983).

22. Bertina, R.M., Broekmans, A.W., Krommenhoek Van, E.S.C., Van Wijngaarden, A.: Thromb. Haemostas. 51, 1-5 (1984).

23. Comp, P.C., Nixon, R.R., Esmon, C.T.: Blood 63, 15-21 (1984).

24. Sala, N., Owen, W.G., Collen, D.: Blood 63, 671-675 (1984).

FUNCTIONAL ASSAYS OF PLASMA PROTEIN C

Rogier M. Bertina and André W. Broekmans
Haemostasis and Thrombosis Research Unit, University Hospital, Leiden, The Netherlands

Introduction

During the past year 4 different methods for the assay of protein C in plasma have been reported (1-6). All assays differ with respect to the functional aspects of protein C that are evaluated (see Tab. 1). During the development and evaluation of these assays it is important to realize when protein C activity measurements will be used. In general, this will be in situations where protein C antigen levels are expected to be less informative, such as analysis of patients for the presence of hereditary protein C deficiency type II (heterozygotes for an abnormal protein C molecule) or analysis of patients with DIC or related disorders, where the turnover of protein C might lead to the presence of significant concentrations of APC-protein C inhibitor complexes. In practice, this means that the treatment of patients with oral anticogulants or heparin should not interfere with the interpretation of the result of the protein C activity assay. In the following paragraphs we will shortly discuss the characteristics of the various protein C activity assays reported so far and finally give some examples of situations where a discrepancy between protein C activity and protein C antigen can be observed.

Protein C - Biochemical and Medical Aspects

Table 1. Functional properties of protein C as evaluated in different protein C activity assays

	Francis & Patch (1)	Bertina et al (6)	Sala et al (4)	Comp et al (2)
Ca^{2+} binding	+	+	+	-
cleavage by thrombin	+	+	+	+
affinity for thrombin	?	+	-	-
affinity for thrombomodulin	-	-	+	+
affinity for protein S	+	-	-	-
presence of active site	+	+	+	+
activity towards				
- factor Va	+	-	-	-
- factor VIIIa	+	-	-	-
profibrinolytic effects	-	-	-	-

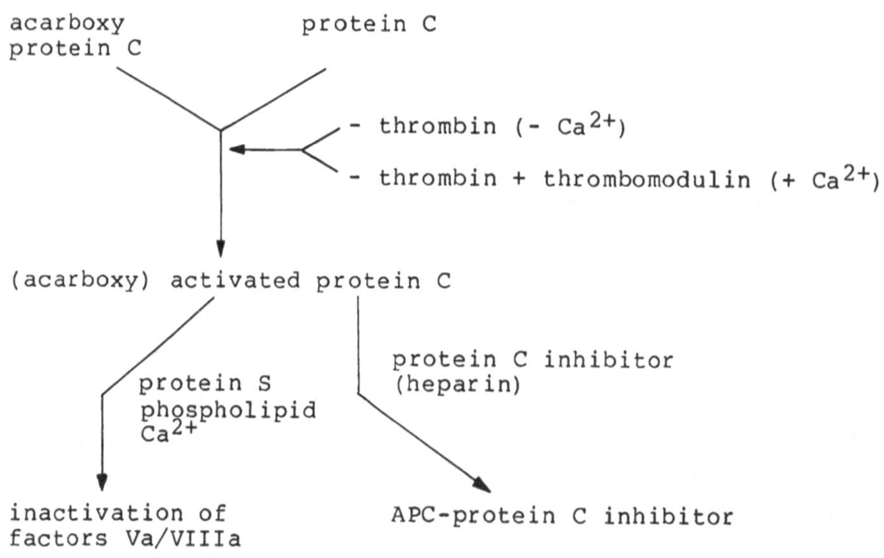

Fig. 1. Functional aspects of protein C

Assay of protein C activity in plasma

Fig. 1 shows schematically the various functional aspects of protein C that are of importance for the evaluation of a protein C activity assay. Important features are: the binding of Ca^{2+} ions, the cleavage by thrombin, the affinity for thrombin and thrombin-thrombomodulin, the affinity of activated protein C for protein S, the presence of an intact active site, and the reactivity of activated protein C towards factor Va and factor VIIIa. Moreover, it should be realized that acarboxy-protein C or heparin might be present, and that there is a specific inhibitor of activated protein C in plasma (7).

Protein C activity assay according to Francis and Patch. The various steps involved in the protein C activity assay according to Francis and Patch (1) are summarized in Tab. 2. Protein C is separated from acarboxy-protein C and the protein C inhibitor via barium citrate adsorption and elution; the eluate - after dilution - is activated with thrombin (no complete activation) after which the thrombin is neutralized with antithrombin III plus heparin; subsequently the heparin

Table 2. Protein C activity assay according to Francis and Patch (1)

- plasma
- barium citrate adsorption and elution (0.2 M MES, 0.15 M citrate, pH 6.0)
- activation with thrombin (partial)
- neutralization of thrombin with antithrombin III plus heparin
- neutralization of heparin with protamine sulphate
- assay of APC (coagulation assay)

is neutralized with protamine sulphate. The concentration of activated protein C in the sample is then compared with that in dilutions of an identically processed pooled normal plasma in a coagulation assay for activated protein C. The advantages of this assay are that it evaluates many of the functional aspects of protein C (see Tab. 1) and that it does not require expensive or not readily available reagents. Disadvantages are the lack of sensitivity of the assay at low protein C concentrations and the limited evaluation of its applicability in patients plasmas (patients with DIC, liver disease, or treated with oral anticoagulants).

Protein C activity assay according to Sala, Owen and Collen. Tab. 3 summarizes the different steps involved in the protein C activity assay developed by Sala et al (4). After the initial adsorption with barium citrate, the eluate is treated with PEG-6000 (to remove the EDTA) and subsequently activated with thrombin-thrombomodulin (complete activation). After neutralization of the thrombin with antithrombin III plus heparin, the amount of activated protein C in the eluate is

Table 3. Protein C activity assay according to Sala, Owen and Collen (4)

- plasma
- barium citrate adsorption and elution (0.1 M EDTA, pH 6.0)
- removal of EDTA (PEG-6000 precipitation)
- activation with thrombin-thrombomodulin
- neutralization of thrombin with antithrombin III plus heparin
- assay of APC (amidolytic, S-2266)

measured using an amidolytic assay with S-2266. For the construction of a calibration curve dilutions of pooled normal plasma in protein C deficient plasma are used. This assay evaluates less functional properties of protein C than the assay described by Francis and Patch (see Tab. 1). However, it includes the important parameter of the reactivity of activated protein C towards thrombin-thrombomodulin. Disadvantages of this assay are that it is a rather laborious assay, that thrombomodulin is required and that it is not known whether this assay is sensitive to the presence of heparin.

Protein C activity assay according to Bertina, Broekmans, Krommenhoek-van Es and Van Wijngaarden. Tab. 4 summarizes the different steps involved in the protein C activity assay developed by Bertina et al (6). They use an adsorption with $Al(OH)_3$ instead of barium citrate for the separation of protein C from acarboxy protein C and the protein C inhibitor. Protein C in the eluate is activated with a relatively high thrombin concentration in the absence of thrombomodulin. Under these conditions the activation of protein C is linear with time and with the protein C concentration (K_m of protein C for the activation by thrombin is very high,

Table 4. Protein C activity assay according to Bertina, Broekmans, Krommenhoek-van Es, Van Wijngaarden (6)

- plasma
- $Al(OH)_3$ adsorption and elution (0.25 M sodium phosphate, pH 7.0)
- activation with thrombin (partial)
- neutralization of thrombin with antithrombin III plus heparin
- assay of APC (amidolytic, S-2366)

cf ref. 8). During the assay only 6 % of the protein C is activated. Thrombin is neutralized with antithrombin III plus heparin, after which the concentration of activated protein C is measured by an amidolytic assay using S-2366. The calibration curve is constructed using dilutions of pooled normal plasma in protein C deficient plasma. This assay does not evaluate the reactivity of protein C towards the thrombin-thrombomodulin complex but gives information on the affinity of protein C for its activation by thrombin. The assay is very sensitive to the presence of low heparin concentrations in the plasma, but this can be overcome efficiently by previous neutralization of the heparin with protamine chloride.

The protein C activity assay according to Comp, Nixon and Esmon. Tab. 5 summarizes the different steps involved in the protein C activity assay developed by Comp et al (2). A slight modification of this assay also has been used by Griffin (personal communication). The design of this assay is quite different from that of the previous three assays. $CaCl_2$ and thrombin-thrombomodulin complex are added to plasma; after a short incubation period antithrombin III is added to neutralize the thrombin. Activated protein C (complete activation) is specifically removed from the plasma by

Table 5. Protein C activity assay according to Comp, Nixon and Esmon

- plasma
- addition of $CaCl_2$, thrombin-thrombomodulin
- neutralization of thrombin with antithrombin III
- immunoadsorption of APC/protein C with anti-protein C IgG-agarose
- washing of immunoadsorption beads
- assay of APC (amidolytic, S-2238)

adsorption with anti-protein C IgG coupled to agarose (DFP-treated!); after washing of the immunoadsorption beads, APC activity (still measurable in the immune complexes) is evaluated with an amidolytic assay using S-2238. One of the disadvantages of this assay is that some of the reagents are not easily available even to specialized laboratories (thrombomodulin, anti-protein C IgG, DFP). An advantage (or disadvantage) is that acarboxy protein C is co-assayed in this procedure, because, in vitro, thrombin-thrombomodulin cleaves protein C and acarboxy protein C equally efficient (9). Moreover, the assay is sensitive to the presence of heparin, a disadvantage which cannot be solved by the previous addition of protamine chloride. The calibration curve is constructed by using dilutions of pooled normal plasma in buffer.

General comments. A common feature of all four protein C activity assays is that these assays are not readily accessible for application in routine laboratories, partly because of the requirement of specific - not easily available - reagents (thrombomodulin, protein C deficient plasma, thrombin, antithrombin III) and partly because of the limited number of samples that can be processed on one day. In general, more information is needed on the sensitivity of these assays to the presence of heparin. Moreover it is recommended to evaluate in greater detail to what degree the result of the assay is dependent on the actual dilution of the plasma sample (especially at high protein C concentrations). Further, it seems highly useful to include a control plasma (preferentially from patients on coumarin-treatment) in each series of assays, to monitor the quality of the assay (a.o. efficiency of the barium citrate or $Al(OH)_3$ adsorption and elution). In general, there is good agreement between the different assays with respect to the range of protein C activity levels in healthy individuals (1-6).

Discrepancy between protein C activity and protein C antigen

A discrepancy between protein C activity and protein C antigen can be expected in various situations as a) hereditary protein C deficiency type II, b) homozygotes for a protein C polymorphism, c) DIC or related disorders, d) oral anticoagulation and e) presence of heparin-like substances.

Hereditary protein C deficiency type II. Several laboratories now have reported patients with thrombotic disease, with normal protein C antigen levels but low protein C activity levels (2, 5, 10). Tab. 6 shows some examples of patients with a type II deficiency, that have been identified in our laboratory. It should be stressed that it is very important to establish the hereditary nature of the defect in this type of deficiency.

Table 6. Hereditary protein C deficiency, type II

	protein C activity (%)	protein C antigen (%)	FIIAg (%)	FXAg (%)
Ib EV	41	110	110	92
VB	47	99	115	135
XIV FS	42	100	125	100
XVI S	35	100	78	80
XV B	38	100	85	89
XXX DV	66	108	99	140

Homozygote for a protein C polymorphism. At present it has been well-established that within the normal population the protein C activity/protein C antigen ratio may vary from 0.64 to 1.57 (6). In this respect it is of great interest to learn whether normals with a low protein C activity/protein C antigen ratio are at risk for the development of venous thrombosis. It is our experience that in some patients an apparent type II protein C deficiency exists, which cannot be confirmed by pedigree analysis. In such a situation, one should be aware of the possibility of homozygosity for a naturally occurring protein C polymorphism.

DIC or related disorders. During disorders in which activation of the coagulation system is significant, one should be aware of the possibility that activated protein C is formed and trapped in complexes with the protein C inhibitor. This can lead to situations in which the protein C activity is much lower than the protein C antigen (see Tab. 7).

Table 7. Protein C activity and antigen in patients with DIC

	protein C activity (%)	protein C antigen (%)
promyelocytic leukemia	39	84
SLE/sepsis	60	105
acute leukemia	43	76

Oral anticoagulation. Because protein C is a so-called vitamin K-dependent coagulation factor, it is expected that its concentration drops during the treatment with vitamin K antagonists. This has been clearly established now for the protein C antigen levels (12, 13, 10). However, because of the presence of non-carboxylated and partially-carboxylated protein C molecules in the circulation, it is expected that only part of the protein C molecules have full protein C activity. This is demonstrated in Tab. 8, in which protein C activity and protein C antigen were compared in plasmas of patients, treated at different intensities of oral anticoagulation (expressed in International Normalized Ratios). However, it is important to realize that the results of such comparisons are dependent on the protein C activity assay used, and even on the actual assay conditions (co-adsorption and elution of partially carboxylated protein C molecules). For the laboratory diagnosis of protein C deficiency in patients using oral anticoagulants, each laboratory should generate its own reference data, obtained with the protein C activity assay, routinely used.

Table 8. Effect of oral anticoagulation on protein C activity and protein C antigen levels

N	INR	protein C activity (%)	protein C antigen (%)
10	2.4 - 2.8	41 ± 5	42 ± 6
10	2.8 - 3.2	38 ± 9	45 ± 8
10	3.2 - 3.6	38 ± 13	50 ± 9
10	3.6 - 4.0	32 ± 5	48 ± 6

Presence of heparin-like substances. As previously mentioned, the presence of heparin in a patient's plasma can interfere in different ways with the protein C activity assay; such as acceleration of the APC-protein C inhibitor interaction (7) and co-adsorption and elution in the barium citrate or $Al(OH)_3$ adsorption step. In the protein C activity assay, developed in our laboratory, this problem can be overcome by the addition of excess protamine chloride to the plasma, prior to analysis. This is demonstrated in the experiment summarized in Tab. 9. Even the presence of 0.1 U/ml heparin in the plasma would lead to an apparently decreased protein C activity level (and thus to a discrepancy with the protein C antigen level). However, the addition of protamine chloride effectively overcomes this unwarranted side effect of heparin.

Table 9. Effect of heparin and protamine chloride on the protein C activity assay

		A 405	(%)
normal plasma		0.592	(100)
protein C deficient plasma		0.112	(0)
normal plasma			
+ heparin:	3.2 U/ml	0.441	(60)
	1.6	0.482	(70)
	0.8	0.448	(65)
	0.4	0.484	(71)
	0.2	0.500	(75)
	0.1	0.467	(66)
+ protamine chloride	5 U/ml	0.587	(96)
+ heparin	2 U/ml		
+ protamine chloride	5 U/ml	0.624	(105)

References

1. Francis, R.B., Patch, M.J.: Thromb. Res. 32, 605-613 (1983).
2. Comp, P.C., Nixon, R.R., Esmon, C.T.: Blood 63, 15-22 (1984).
3. Sala, N., Owen, W.G., Collen, D.: Thromb. Haemostas. 50, 352 (1983).
4. Sala, N., Owen, W.G., Collen, D.: Blood 63, 671-675 (1984).
5. Bertina, R.M., Broekmans, A.W., Krommenhoek-van Es, C., Van Wijngaarden, A.: Thromb. Haemostas. 50, 350 (1983).
6. Bertina, R.M., Broekmans, A.W., Krommenhoek-van Es, C., Van Wijngaarden, A.: Thromb. Haemostas. 51, 1-5 (1984).
7. Suzuki, K., Nishioka, J., Hashimoto, S.: J. Biol. Chem. 258, 163-168 (1983).
8. Owen, W.G., Esmon, C.T.: J. Biol. Chem. 256, 5532-5535 (1981).

9. Esmon, N.L., Debault, L.E., Esmon, C.T.: J. Biol. Chem. 258, 5548-5553 (1983).

10. Griffin, J.H.: Semim. Thromb. Haemostas. 10, 162-166 (1984).

11. Barbui, T., Finazzi, G., Mussoni, L., Riganti, M., Donati, M.B., Colucci, M., Collen, D.: The Lancet II 8406, 819 (1984).

12. Griffin, H.J., Evatt, B., Zimmerman, T.S., Kleiss, A.J., Wideman, C.: J. Clin. Invest. 68, 1370-1373 (1981).

13. Bertina, R.M., Broekmans, A.W., van der Linden, I.K., Mertens, K.: Thromb. Haemostas. 48, 1-5 (1982).

HEREDITARY PROTEIN C DEFICIENCY

André W. Broekmans and Rogier M. Bertina
Haemostasis and Thrombosis Research Unit, University Hospital, Leiden, The Netherlands

Introduction

Since the first report of Griffin et al. (1) in 1981 on hereditary protein C deficiency as risk factor for the development of thrombotic disease, many patients with protein C deficiency have been identified both in Europe and in the United States of America (2, 3, 4, 5, 6, 7). The current literature clearly demonstrates that in some respects the clinical manifestations of hereditary protein C deficiency differ from what has been reported on antithrombin III deficiency (8). This review will summarize present available information on diagnosis, clinical manifestations, prevalence and treatment of hereditary protein C deficiency.

Laboratory diagnosis

Two types of hereditary protein C deficiency can be recognized (Tab. 1; ref. 9). In type I protein C deficiency there is a defect in one of the genes coding for protein C that will not permit translation into a protein C-like molecule. In these patients the plasma concentration of protein C activity and/or protein C antigen will be below the lower limit of the normal range (Tab. 1, 2), while the ratio between protein C activity and protein C antigen still is within the normal range (9).

Protein C - Biochemical and Medical Aspects

Table 1. Criteria for the laboratory diagnosis of protein C deficiency

	Type I	Type II
Protein C activity	↓*	↓*
Protein C antigen	↓*	normal
Ratio protein C activity/protein C antigen	normal	↓*
Factors VII, IX, X, II	normal	normal
Disseminated intravascular coagulation	absent	absent
Chronic liver disease	absent	absent
Oral anticoagulant treatment	stable	stable

↓*: below the lower limit of the normal range

Table 2. Normal values of protein C

	mean $\pm$ SD	range
Protein C activity (U/ml)	0.94 $\pm$ 0.17	0.61 - 1.32
Protein C antigen (U/ml)	0.99 $\pm$ 0.17	0.67 - 1.40
Ratio protein C activity/ protein C antigen	0.97 $\pm$ 0.21	0.64 - 1.57

In type II protein C deficiency one of the genes coding for protein C has been modified in such a way that an abnormal gene-product is produced. In the laboratory these defective molecules can only be detected by the combined use of a functional and an immunologic assay for protein C. In these patients protein C antigen will be within the normal range, while the protein C activity and the ratio between protein C activity and protein C antigen are below the lower limit of the normal range (Tab. 1, 2; ref. 9).

In both types of protein C deficiency the plasma concentration of the other vitamin K-dependent coagulation factors should be within the normal range. The presence of conditions known to be associated with acquired protein C deficiency, such as unstable oral anticoagulant treatment, vitamin K deficiency, liver disease and disseminated intravascular coagulation (2, 10, 11, 12) should be excluded.

During oral anticoagulant treatment the concentrations of the vitamin K-dependent factors are decreased. The actual reduction of the concentration of the vitamin K-dependent factors is a function of the intensity of oral anticoagulant treatment (2, 9). Because many patients suffering from thrombotic disease will be treated with oral anticoagulants, it is important to establish normal ranges for each of the vitamin K-dependent factors (activity and antigen) at varying intensities of treatment in order to detect an isolated deficiency in one of these factors. In our laboratory reference values have been established for protein C antigen, ratios between protein C antigen and factor II antigen and between protein C antigen and factor X antigen at increasing intensities of oral anticoagulant treatment (2). These reference values can only be used when the patient to be studied is in a stable phase of oral anticoagulant treatment at the time of the investigation, because during unstable oral anticoagulant treatment considerable differences in the concentrations of vitamin K-dependent factors can occur due to differences in biological half-lifes (2, 12). This is illustrated in Tab. 3 by the laboratory findings in a female patient who was studied at three different occasions. The first time, she was studied three days after the initiation of oral anticoagulant treatment and was found to be protein C deficient. However, after the anticoagulant therapy had been stabilized, no isolated protein C deficiency could be detected.

Table 3 E.K.-V., born 18-08-1946
diagnosis: pulmonary embolism; no family history
19-11-1982: start of oral anticoagulant treatment

	22-11	01-12	08-12
protein C antigen (U/ml)	0.56	0.50	0.65
factor II antigen (U/ml)	1.05	0.64	0.66
factor X antigen (U/ml)	0.96	0.50	0.48
Thrombotest (%)	14	8	9

Heterogeneity of protein C deficiency

The first reports on families with hereditary protein C deficiency all concerned type I protein C deficiencies (1, 2, 3, 4, 5, 7). As is illustrated in Fig. 1, type I protein C deficiency is inherited as an autosomal dominant trait (3, 4).

At present in The Netherlands 52 patients from 24 different families have been identified with a type I protein C deficiency. All these patients are heterozygous for the defect as is reflected in an average plasma concentration of about 0.50 U/ml for both protein C activity and protein C antigen (Tab. 4). However, individual protein C values may vary between 0.30 U/ml and 0.76 U/ml. This wide range of protein C concentrations in heterozygotes is compatible with the variable expressivity of the protein C gene (3).

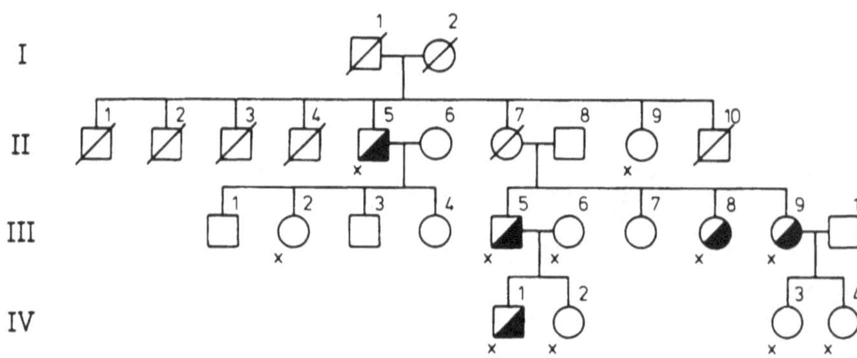

Fig. 1. Autosomal dominant inheritance pattern of protein C deficiency

 Heterozygotes for protein C deficiency

X Investigated

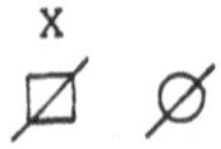 Deceased

Table 4: Laboratory values in type I protein C deficient patients (n = 52)

	- OAC[1]	+ OAC
Protein C activity (U/ml)	0.51 ± 0.11[2] (0.30 - 0.74)[3]	0.12 ± 0.08 (0.05 - 0.23)
Protein C antigen (U/ml)	0.50 ± 0.10 (0.31 - 0.76)	0.23 ± 0.05 (0.13 - 0.29)

1 OAC = oral anticoagulant treatment

2 mean ± SD

3 range

Three homozygous patients with type I protein C deficiency have been reported (5, 13, 14). In all cases it concerned children from consanguinous marriages with both parents being heterozygous for protein C deficiency. Two of these homozygous patients had a severe defect (protein C antigen concentration < 0.01 U/ml) and died in the neonatal period from extensive thrombosis (5, 13). The other reported case of homozygous protein C deficiency had a mild defect (protein C antigen 0.24 U/ml) and had his first thrombotic episode at the age of 17 years (14).
At present only four patients with a type II protein C deficiency have been reported (9, 15).

Clinical manifestations of hereditary protein C deficiency

There is evidence from many centres that patients heterozygous for protein C deficiency are at risk for the development of venous thrombo-embolism (1, 3, 4, 5, 7, 16). An interesting family has been reported by Seligsohn et al (5); in this family only the homozygous proband but none of the 14 heterozygotes had thrombotic disease. Further studies should

reveal whether other factors are operational in these heterozygotes to protect them against the occurrence of venous thrombo-embolism.
The following clinical manifestations can be recognized:

Venous thrombo-embolism

The majority of patients with a hereditary protein C deficiency have a history of thrombotic disease (1, 3, 4, 7). A protein C concentration of 0.65 U/ml or less evidently constitutes a distinct risk for the development of thrombotic disease (3). In The Netherlands 77 % of the patients with a type I protein C deficiency have a history of venous thrombo-embolism (Tab. 5), often with recurrences (in 70 % of the patients with thrombotic disease). Deep venous thrombosis

Table 5. Clinical manifestations in type I protein C deficient patients (n = 52)

	n	%
no venous thrombo-embolism	12	23
superficial thrombophlebitis	25	48
deep venous thrombosis	33	63
pulmonary embolism	21	40
recurrences	28	54

was present in 63 % of the patients, often in combination with pulmonary embolism (in 40 % of the patients). Two patients suffered from axillary vein thrombosis. In the Dutch patient material the frequency of superficial thrombophlebitis is high (48 %), in 10 % of the patients it was even the only manifestation of thrombotic disease. This high incidence of superficial thrombophlebitis is not confirmed by all authors (4, 7). Patients may also be at risk for the development of mesenterial vein thrombosis (4). Mostly the first thrombotic episode occurs without an apparent cause (67 % of the Dutch patients), but also known risk factors as

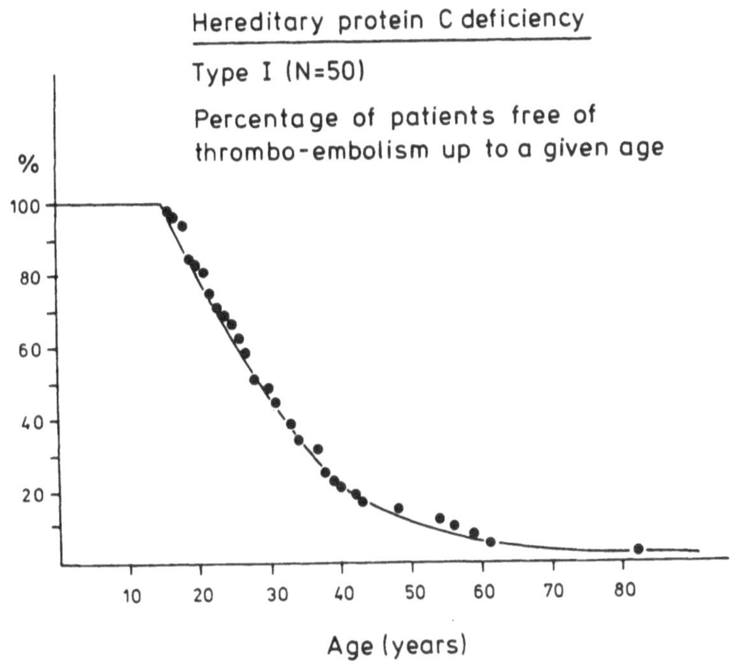

Fig. 2. Percentage of patients free of venous thrombo-embolism up to a given age

surgery, immobilization, and pregnancy were present (3, 4, 7). Fig. 2 shows the percentage of patients free of thrombo-embolism as a function of age. In the Dutch patient material about 50 % of all patients had at least one thrombotic episode before the age of 30 years and 80 % before the age of 40 years. None of the Dutch patients had clinical manifestations before the age of 15 years.

<u>Cerebral haemorrhagic infarction due to venous thrombosis</u>

Three female patients with a type I protein C deficiency suffered at a relatively young age (27, 34 and 38 years of age, respectively) from a cerebral event attributed to cerebral venous thrombosis (17). In all cases the cerebral event was preceded by a period of prodromal headache. A cerebral haemorrhagic infarction was documented by CT scan in two patients and at autopsy in the patient who died. All three female patients had a history of venous thrombo-embolism and there was no evidence for arterial thrombo-embolism as cause for the cerebral incident. In one patient the available angiogram showed cerebral venous thrombosis (17).

Thrombosis in the microcirculation

The development of coumarin-induced haemorrhagic skin necrosis has been reported in 4 patients with hereditary protein C deficiency (18, 14, 19). Three patients were heterozygous for the defect and the fourth patient was a homozygote. The repeated occurrence of coumarin-induced haemorrhagic skin necrosis seems to be pathognomonic for the presence of hereditary protein C deficiency with a severe defect (14, 18). Haemorrhagic skin necrosis is secondary to diffuse thrombosis in the venules and interstitial bleeding (20). The lesions occur predominantly in the initial phase of oral anticoagulant treatment (20) but are also observed in vitamin K deficiency due to cholestasis (21). During the initial phase of oral anticoagulant therapy there is a rapid drop in factor VII and in protein C, while the factors IX, X and II decrease at a slower rate (12). Such a transient state of acquired protein C deficiency might constitute a risk for thrombosis, especially when thrombin can still be generated via the intrinsic - factor VII independent - pathway (18). This risk will be even higher in patients with hereditary protein C deficiency.
In three infants with severe protein C deficiency a syndrome resembling purpura fulminans has been described (22, 13, 23). Biopsy of the affected skin lesions showed thrombosis of virtually all superficial and deep vascular structures with fibrinoid necrosis of the vessel wall (13). Post-mortem examination of the homozygous newborn described by Seligsohn

and co-workers also demonstrated thrombosis in the microcirculation in addition to deep venous thrombosis and pulmonary embolism (5).
The purpura fulminans syndrome in severe protein C deficiency and the coumarin-induced haemorrhagic skin necrosis may represent the same pathogenetic entity.

Lack of arterial thrombotic disease

Only a few protein C deficient patients have been reported with symptomatic arterial thrombotic disease (1, 6). Not all authors confirm the presence of arterial thrombosis (3, 4, 7). In the Dutch patients with type I protein C deficiency, none had evidence of arterial thrombotic disease at an age younger than 50 years.

Incidence of hereditary protein C deficiency

No information is available yet on the incidence of protein C deficiency in the general population. Most studies have been performed in groups of patients with venous thrombo-embolism referred to a specialized centre. In Austria 2 patients with type I protein C deficiency were detected in a group of 225 patients with a history of venous thrombo-embolism (4). Horellou and co-workers (7) discovered in France 9 patients with type I protein C deficiency among 118 patients with documented thrombotic disease. On the other hand, Jespersen and Bertina (24) did not detect one single patient with protein C deficiency in a group of 110 patients with thrombotic disease.
In Leiden the prevalence of protein C deficiency was estimated in an unselected group of patients (25). Two groups of patients were selected out of all patients (n = 800) treated for venous thrombo-embolism with vitamin K antagonists at May 1, 1982. In the first group all patients younger than 41 years of age (n = 62) treated for either superficial thrombo-

phlebitis, deep venous thrombosis and/or pulmonary embolism were included, while the second group consisted of all patients older than 40 years of age (n = 257) treated for recurrence of venous thrombo-embolism. The prevalence of type I protein C deficiency in patients younger than 41 years of age was found to be 4.8 %. The frequency of type I protein C deficiency in all patients treated for recurrent thrombotic episodes was 2.2 % and even higher (13.6 %) in the group of patients younger than 41 years of age.

Treatment of hereditary protein C deficiency

Long-term treatment with coumarin drugs effectively prevents recurrent venous thrombo-embolism in heterozygotes (3, 7). This is supported by the observation that patients may suffer from recurrences of thrombotic disease shortly after withdrawal of the coumarin drug and that they remain asymptomatic during long-term treatment (3, 7). However, in the initial phase of oral anticoagulant treatment patients are at risk for the development of haemorrhagic skin necrosis (14, 18 19). Co-administration of heparin during the initial phase does not always prevent the occurrence of skin necrosis (14, 18, 19). Skin necrosis may be avoided when oral anticoagulant treatment is started with a low dose of the drug and the prothrombin time is regularly monitored with a factor VII-sensitive thromboplastin.
According to Pabinger-Fasching et al. (4), heparin is, however, effective for the prevention of venous thrombo-embolism in situations like surgery.
In case of severe (often homozygous) protein C deficiency with extensive venous thrombosis substitution therapy is indicated with fresh frozen plasma or prothrombin complex concentrates (22, 23) until the effect of oral anticoagulant therapy has been stabilized. In this respect it is important to realize that heparin is also not able to prevent the

occurrence of thrombosis in the microcirculation (22). Recently, it has been reported that stanozolol, an anabolic steroid, raises the concentration of plasma protein C antigen with 50 to 60 % in healthy individuals (26). We treated a type I protein C deficient patient with stanozolol during 6 weeks. As can be seen in Fig. 3, the protein C antigen concentration increased from 46 % to 68 %. Additional studies in protein C deficient patients are needed to establish the significance of this observation and to assess the efficacy of this treatment. However, it should be recognized that stanozolol may have serious side-effects and may potentiate the activity of coumarin drugs (27).

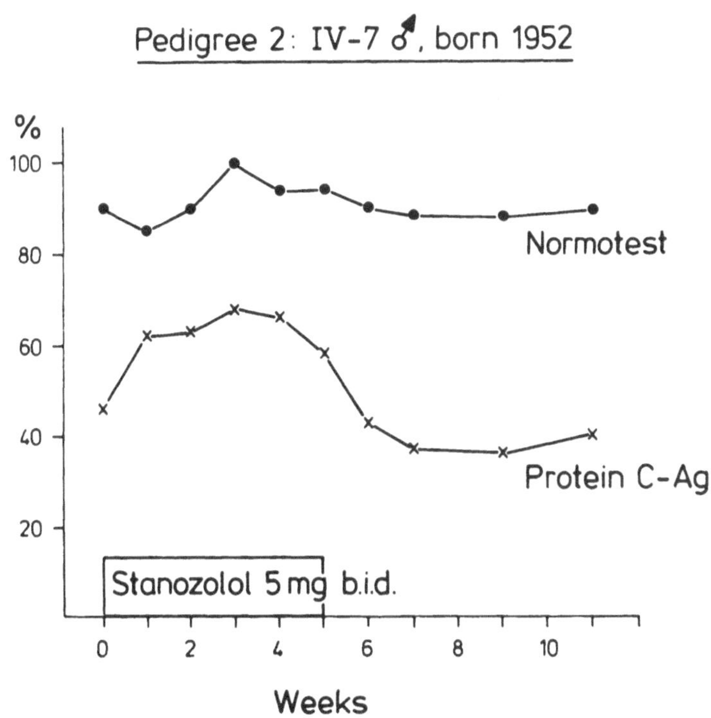

Fig. 3. Treatment with stanozolol of a patient with type I protein C deficiency

References

1. Griffin, J.H., Evatt, B., Zimmerman, T.S., Kleiss, A.J., Wideman, C.: J. Clin. Invest. 68, 1370-1373 (1981).

2. Bertina, R.M., Broekmans, A.W., Van der Linden, I.K., Mertens, K.: Thromb. Haemostas. 48, 1-5 (1982).

3. Broekmans, A.W., Veltkamp J.J., Bertina R.M.: N. Engl. J. Med. 309, 340-344 (1983).

4. Pabinger-Fasching, I., Bertina, R.M., Lechner, K., Niessner, H., Korninger, Ch.: Thromb. Haemostas. 50, 810-813 (1983).

5. Seligsohn, U., Berger, A., Abend, M., Rubin, L., Attias, D., Zivelin, A., Rapaport, S.I.: N. Engl. J. Med. 310, 559-562 (1984).

6. Marlar, R.A., Endres-Brooks, J.: Thromb. Haemostas. 50, 351 (1983).

7. Horellou, M.H., Conard, J., Bertina, R.M., Samama, M.: Br. Med. J. 289, 1285-1287 (1984).

8. Thaler, E., Lechner, K.: Clin. Haematol. 10, 369-390 (1981).

9. Bertina, R.M., Broekmans A.W., Krommenhoek-van Es, C., Van Wijngaarden, A.: Thromb. Haemostas. 51, 1-5 (1984).

10. Mannucci, P.M., Vigano, S.: Lancet ii, 463-466 (1982).

11. Griffin, J.H., Mosher, D.F., Zimmerman, T.S., Kleiss, A.J.: Blood 60, 261-264 (1982).

12. Vigano, S., Mannucci, P.M., Solinas, S., Bottasso, B., Mariani, G.: Br. J. Haematol. 57, 213-220 (1984).

13. Marciniak, E., Wilson, H.D., Marlar, R.A.: Blood 62, suppl. 303 a (1983).

14. Samama, M., Horellou, M.H., Soria, J., Conard, J., Nicolas, G.: Thromb. Haemostas. 51, 132-133 (1984).

15. Comp, P.C., Nixon, R.R., Esmon, C.T.: Blood 63, 15-21 (1984).

16. Bercoff, E., Morcamp, D., Bourreille, J., Borg, J.Y., Piguet, H., Soria, J., Soria, C., Dorner, M.: Presse Méd. 13, 49 (1984).

17. Wintzen, A.R., Broekmans, A.W., Bertina, R.M., Briet, E., Briet, P.E., Zecha, A., Vielvoye, G.J., Bots, G.Th.A.M.: Br. Med. J., accepted for publication.

18. Broekmans, A.W., Bertina, R.M., Loeliger, E.A., Hofmann, V., Klingemann, H.G.: Thromb. Haemostas. 49, 244 (1983).

19. McGehee, W.G., Klotz, T.A., Epstein, D.J., Rapaport, S.I.: Ann. Intern. Med. 100, 59-60 (1984).

20. Loeliger, E.A., Broekmans, A.W. In: Dukes MNG, ed. Meyler's side effects of drugs. Amsterdam, Oxford, Princeton: Excerpta Medica, 10th ed. 648-693 (1984).

21. Hofmann, V., Frick, P.G.: Thromb. Haemostas. 48, 245-246 (1982).

22. Branson, H.E., Katz, J., Marble, R., Griffin, J.H.: Lancet ii, 1165-1168 (1983).

23. Marlar, R.A., Sills, R.H., Montgomery, R.R.: Blood 62, suppl. 303 a (1983).

24. Jespersen, J., Bertina, R.M. Unpublished observations (1983).

25. Broekmans, A.W., Van der Linden, I.K., Veltkamp, J.J., Bertina, R.M.: Thromb. Haemostas. 50, 350 (1983).

26. Kluft, C., Bertina, R.M., Preston, F.E., Malia, R.G., Blamey, S.L., Lowe, G.D.O., Forbes, C.D.: Thromb. Res. 33, 297-304 (1984).

27. Jarrett, P.E.M., Morland, M., Browse, N.L. In: Davidson J.F., ed. Progress in chemical fibrinolysis and thrombolysis. Edinburgh: Churchill Livingstone 317-321 (1979).

PRODUCTION OF AN ANTIBODY AGAINST HUMAN PROTEIN C AND ITS USE TO DETECT PATIENTS WITH CONGENITAL PROTEIN C DEFICIENCY ASSOCIATED WITH THROMBOSIS

Jean-Marie Freyssinet, Lélia Grunebaum, Marie-Louise Wiesel, Jean-Pierre Cazenave

Service d'Hémostase et de Thrombose, Centre Régional de Transfusion Sanguine, 10 rue Spielmann, 67085 Strasbourg, France

Rodolphe Fritsché, Susanne Greber, Alfred Lampart

Merz and Dade AG, 3186 Düdingen, Switzerland

Reiner Spaethe

AHS Deutschland, Department Merz and Dade, München, FRG

Human protein C (PC) is a vitamin K-dependent zymogen with a molecular weight of 62,000 which circulates in plasma at a concentration of about 3 µg/ml (1, 2). The only known physiological activator of PC is alpha-thrombin. Thrombin requires the presence of either thrombomodulin, an endothelial cell surface protein (3) or factor Va (4) as cofactor to activate PC. At physiological concentration, calcium is an inhibitor of the conversion of PC into its activated form PCa by thrombin alone (5). PCa is a serine protease which exerts anticoagulant activity by proteolytic cleavage of factors Va and VIIIa (6, 7) in the presence of a cofactor, protein S, also a vitamin K-dependent protein (8). PCa is inactivated by a natural and specific inhibitor present in circulating blood (9). Several clinical studies of families with histories of unexplained, recurrent thrombosis have recently suggested that an inherited deficiency of PC might predispose patients to throm-

Protein C - Biochemical and Medical Aspects

boembolic disease (10). It is possible that other components of the PC system (protein S, thrombomodulin) will prove to be of clinical importance.
In the present study, we describe the large scale isolation of human PC from human cryosupernatant in order to obtain a specific antiserum useful for clinical or research purposes.

Purification of human PC from plasma cryosupernatant

PC was routinely isolated from 25 liters of cryosupernatant following a procedure derived from methods published by Kisiel (1), Suzuki et al. (6) and Freyssinet et al. (11).
The presence of PC was either detected by the ability of its active form, PCa, to prolong activated partial thromboplastin time (APTT) of normal plasma or by rocket electroimmunoassay, when the antiserum became available.
Vitamin K-dependent proteins were adsorbed on barium citrate precipitate. After several washes, the eluted proteins were fractionated successively by 35 % and 65 % saturated ammonium sulfate solutions (1). After dialysis, the protein collected in the last precipitate was subjected to two DEAE-Sephacel ion exchange chromatographies. The first one was performed at pH 6.0 in the presence of EDTA, while the second one was run in the presence of Ca^{2+} at pH 7.5 (6). At this stage, only factor X could be detected by clotting assays. PC and factor X were separated by dextran sulfate-sepharose 4B affinity chromatography. Following conditions described in (11), PC eluted first in a symmetrical peak, then factor X eluted in a symmetrical peak. PC was further purified on the same dextran sulfate-Sepharose 4B affinity chromatography (Fig. 1). PC preparation was greater than 95 % pure, as judged by polyacrylamide gel electrophoresis in the presence of sodium dodecyl sulfate (Fig. 2). As pointed out by other authors (1, 6) most of our preparations contained a small proportion of degraded PC heavy chain (molecular weight about 38,000, instead of 41,000). Purified PC contained less than 1 % factor X,

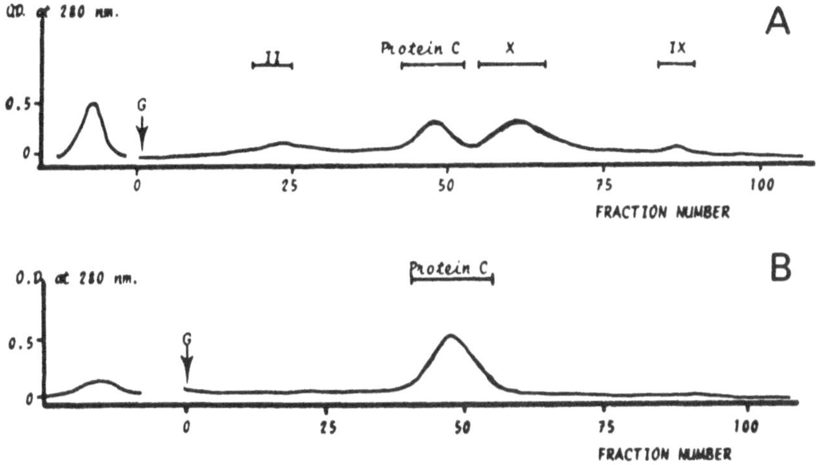

Fig.1. A. Dextran sulfate-Sepharose 4B affinity chromatography of the pool enriched in PC from DEAE-Sephacel chromatographies. Buffer is 5 mM MOPS containing 8 mM trisodium citrate and 0.01 % NaN_3, pH 7.5. Gradient, G, is from 0 to 1 M NaCl (2x400 ml). Volume of the column is 40 ml. Volume of a fraction is 7 ml.

B. Rechromatography of the PC pool obtained in A on dextran sulfate-Sepharose 4B. Conditions are identical to those of A except volume of column is 10 ml and volume of gradient is 2x100 ml. Volume of a fraction is 10 ml and volume of gradient is 2x100 ml. Volume of a fraction is 1.8 ml.

based on clotting assays. The isolation method of PC is summarized in Tab. 1.

Preparation of anti-protein C antiserum

Rabbit anti-human PC antiserum was prepared by immunizing rabbits with four weekly injections of 30 - 50 µg purified PC in Freund's (in)complete adjuvant.

Immune sera were adsorbed first with $BaSO_4$ (100 mg/ml), decomplemented during 30 min at 56°C and then adsorbed once more with 1/10 (v/v) of human plasma previously adsorbed on 5 % (w/v) $Al(OH)_3$ as described in (12). Analysis by Ouchterlony double diffusion and by immunoelectrophoresis have shown that this antiserum contains specific and strongly precipitating antibodies against human PC (Fig. 3).

Table 1. Purification of human PC from 25 liters of cryosupernatant

Step	Volume (%)	Total protein (mg)	Recovery (%)
Cryosupernatant	25×10^3	1.5×10^5	
Barium citrate eluate	250	3.5×10^3	
DEAE-Sephacel-EDTA	200*	210	
DEAE-Sephacel-Ca^{2+}	80*	65	
1st dextran sulfate-Sepharose	60*	10	
2nd dextran sulfate-Sepharose	25*	7	10

*Values before concentration on Amicon PM 10 membrane.

Measurement of PC antigen by rocket electroimmunoassay

The electrophoresis buffer was Tricine at pH 8.6: 17.2 g Tricine (Biorad), 38.2 g Tris, 1 mM EDTA in 4 liters of H_2O. Agarose (Standard Low-mr, Biorad) was used at a final concentration of 1 % (w/v) with anti-PC antiserum present at a final dilution of 1/75. The electrophoresis was carried out for 18 hours at 10 mA/plate. The gels were then washed in TWEEN-NaCl solution (0.9 % (w/v) NaCl and 0.15 % (v/v) TWEEN 20) for 24 hours. The gels were dried, stained with Coomassie Blue and destained in 10 % acetic acid solution. Fig. 4 shows

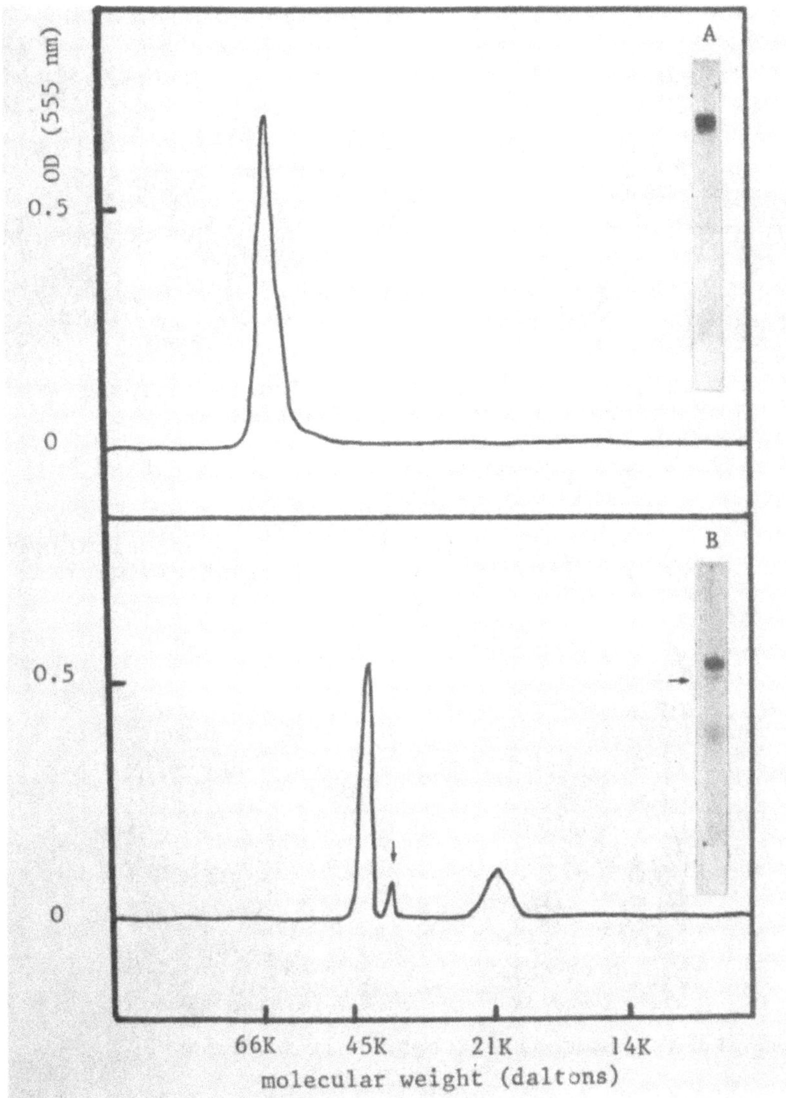

Fig. 2. SDS-polyacrylamide gel electrophoresis of human PC.

A. 8 µg of unreduced PC
B. 8 µg of reduced PC (5 % ß-mercaptoethanol)

The arrows indicate the degraded form of the PC heavy chain.

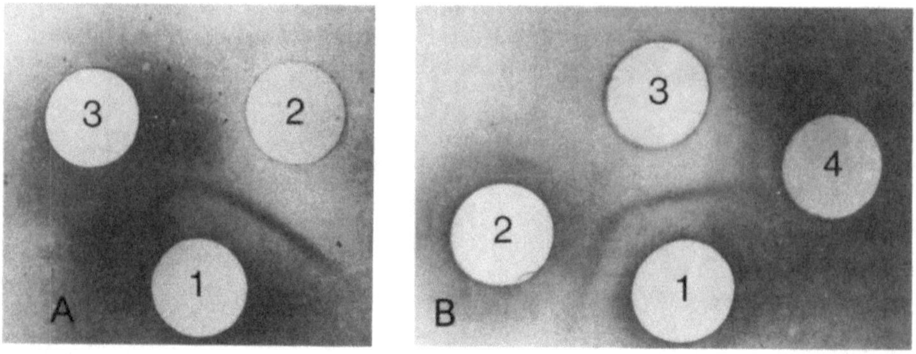

Fig. 3. Ouchterlony double immunodiffusion analysis of the rabbit antihuman PC antiserum.

A. 1 anti-human PC absorbed; 2 purified PC; 3 fresh normal human plasma.

B. 1 anti-hPC absorbed; 2 fresh normal plasma; 3 purified PC; 4 PC deficient plasma.

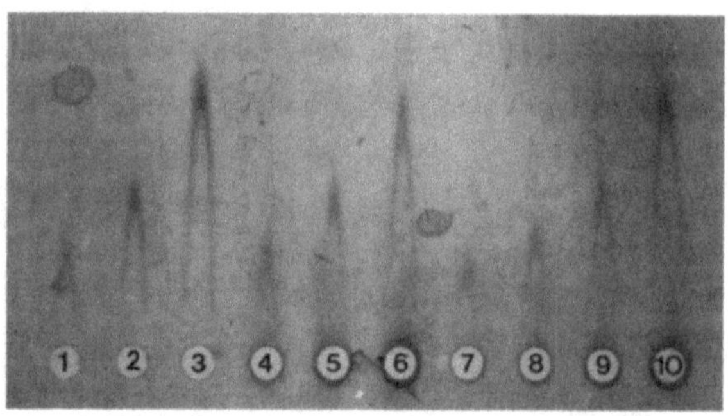

Fig. 4. Rocket immunoelectrophoresis of human PC.

1-3) Purified PC diluted 1/4 (1), 1/2 (2), 1/1 (3).
4-5) Reference plasma (Coag-cal®) diluted 1/4 (4), 1/2 (5), 1/1 (6).
7-10) Fresh normal plasma pool diluted 1/8 (7), 1/4 (8), 1/2 (9), 1/1 (10).

different samples containing PC analyzed by this technique using the above antiserum. The limit of detection of PC in plasma was about 0.08 U/ml.

Measurement of PC activity

In order to measure PC activity, we have used a procedure similar to that described by Bertina et al. (13), except that vitamin K-dependent proteins were adsorbed on barium citrate instead of $Al(OH)_3$. The average recovery of prothrombin was about 80 %. As pointed out by Sala et al. (2), reproducible recoveries of prothrombin were only obtained when the temperature of the barium citrate adsorption step was well controlled.

The activation of PC was carried out by pure human alpha thrombin in the presence of EDTA. When activation was complete, thrombin was inhibited by an excess of hirudin. Generated PCa was then estimated by measuring its amidolytic activity with the chromogenic substrate S-2238 (Kabi). For all measurements, the proteolytic background due to possible activation by thrombin of protein other than PC was subtracted. For this purpose, we have used a PC depleted plasma prepared by immunoadsorption with PC antibodies.

Detection of congenital protein C deficiency in patients with familial recurrent thrombosis

The etiology of familial recurrent thrombosis is a difficult problem for the biologist and the clinician. Until the description by Griffin et al. (14) and Bertina et al. (12) of an association of PC deficiency and congenital thrombotic disease, the other known causes of inherited thrombosis were abnormalities of factor XII, antithrombin III, fibrinogen, plasminogen activator and plasminogen. Since then, many more cases of heterozygous and homozygous congenital PC deficiency have been described all over the world and associated with various clinical manifestations ranging from few or no thrombotic events, frequent recurrent venous thrombo-embolic episodes, arterial thrombosis, antivitamin K-induced skin necrosis to massive thrombosis of the microcirculation with purpura fulminans (A.W. Broekmans, this volume). Congenital PC deficiency seems to be more frequent than antithrombin III deficiency. The prevalence of PC deficiency has been estimated to be one in 16,000.

In a one year period, our laboratory has been screening for isolated PC deficiency in patients with recurrent venous thrombosis and unknown abnormalities of platelets, coagulation and fibrinolytic systems. We have briefly reported the clinical and biological study of a large French family with congenital deficiency of PC (both antigenic and biological activities, type I deficiency) associated with recurrent venous thrombosis (15 and Tab. 2, Family D...). Since then, we have discovered 4 new families with congenital PC deficiency. In these families, PC levels were measured by rocket immunoelectrophoresis with the antibody we have developed. The results are presented in Tab. 2 with the main clinical manifestations. All these 14 patients have mean PC levels about 50 % of the normal value and are probably heterozygous for the trait, which is inherited as autosomal dominant.

Table 2. Patients with familial deficiency of PC.

Patients	Protein C antigen (U/ml)	Thrombotic manifestations
Normal population (range)	0.66-1.2	none
<u>Family D...</u>		
II 2	0.54	varicose veins
II 6	0.49	recurrent superficial venous thrombosis
II 8	0.31	recurrent deep venous thrombosis
II 15	0.45	recurrent deep venous thrombosis
10 normal members (range)	0.87-1.32	none
<u>Family W...</u>		
III 2	0.35	recurrent deep venous thrombosis
III 4	0.42	recurrent deep venous thrombosis
IV 1	0.44	none
IV 2	0.33	none
IV 3	0.52	none
IV 4	0.33	none
<u>Family J...</u>		
II 1	0.59	none
II 2	0.40	none
<u>Family H...</u>		
II 1	0.46	recurrent venous thrombosis
<u>ROH. F...</u>	0.19	recurrent deep venous thrombosis under AVK (PT 44 %)

The level of PC antigen was measured in plasma by rocket immunoelectrophoresis.

References

1. Kisiel, W.: J. Clin. Invest. 64, 761-769 (1979).

2. Sala, N., Owen, W.G., Collen, D.: Blood 63, 671-675 (1984).

3. Esmon, C.T., Esmon, N.L.: Semin. Thromb. Hemostas. 10, 122-130 (1984).

4. Salem, H.H., Broze, G.J., Miletich, J.P., Majerus, P.W.: Proc. Natl. Acad. Sci. USA 80, 1584-1588 (1983).

5. Amphlett, G.W., Kisiel, W., Castellino, F.J.: Biochemistry 20, 2156-2161 (1981).

6. Suzuki, K., Stenflo, J., Dahlbäck, B., Teodorsson, B.: J. Biol. Chem. 258, 1914-1920 (1983).

7. Vehar, G.A., Davie, E.W.: Biochemistry 19, 401-409 (1980).

8. Walker, F.J.: Semin. Thromb. Hemostas. 10, 131-138 (1984).

9. Suzuki, K.: Semin. Thromb. Hemostas. 10, 154-161 (1984).

10. Griffin, J.H.: Semin. Thromb. Hemostas. 10, 162-166 (1984).

11. Freyssinet, J.-M., Thevenon, D., Souque, A., Suscillon, M.: Thromb. Haemostas. 48, 120-124 (1982).

12. Bertina, R.M., Broekmans, A.W., Van der Linden, I.K., Mortens, K.: Thromb. Haemostas. 48 1-5 (1982).

13. Bertina, R.M., Broekmans, A.W., Krommenhoek-van, E., van Wijngarden, A.: Thromb. Haemostas. 51, 1-5 (1984).

14. Griffin, J.H., Evatt, B., Zimmerman, T.S., Kleiss, A.J., Wideman, C.: J. Clin. Invest. 68, 1370-1373 (1981).

15. Cazenave, J.-P., Wiesel, M.-L., Grunebaum, L., Freyssinet, J.-M., Dorner, M., Fritsché, R., Greber, S., Spaethe, R., Lampart, A.: VIIIth International Congress on Thrombosis, Istanbul, June 4-7 (1984) (Abstract).

PROTEIN C DEFICIENCY IN A NEWBORN INFANT WITH PURPURA FULMINANS

Helmut Wehinger and Irene Witt

Kinderklinik der Städtischen Kliniken
D-3500 Kassel
Biochemisches Labor der Universitäts-Kinderklinik
D-7800 Freiburg

Purpura fulminans is a rare hemorrhagic disorder with high mortality. The disease may occur in newborn infants (1, 2). The pathogenesis has not been established definitely. Recently protein C deficiency was diagnosed in a newborn infant with purpura fulminans (3). We report on a similar case and successful treatment with phenprocoumon.

Case history

The child was born on 17th August 1983 after an uneventful pregnancy weighing 2920 g. The father was 22, the mother 21 years old; both were healthy. The maternal grandmother has had one episode of thrombophlebitis in her leg. The newborn baby presented with normal vital signs; Apgar score was 10 after five minutes. There was a swelling and bluish discoloration of the scrotum, and it was impossible to palpate the testes. A diagnosis of intrauterine testicular torsion was made. This diagnosis proved false, when the hematoma was resorbed after a few days and other hematomas developed instead. On the sixth day of life a bluish-black discoloration of the right elbow appeared, on day 11 a necrotizing hemorrhage of the lower abdomen, and on day 14 an extensive bleeding involving the right scalp. There were more relapses and a

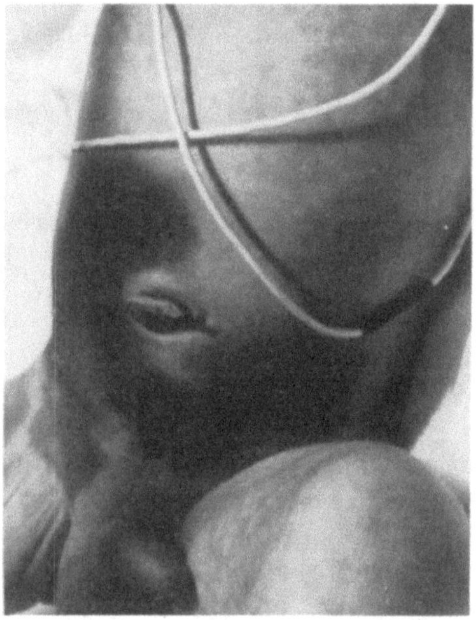

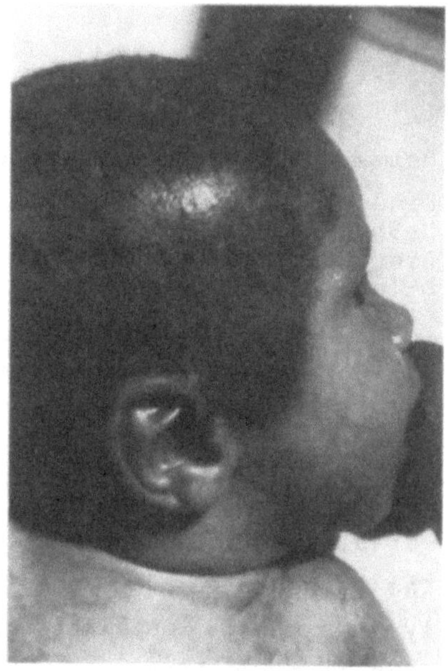

Fig. 1. Large hematoma on the right side of the scalp and on the lower abdomen

total of nine hemorrhagic episodes were observed, most of them occurring on the scalp. All these lesions were sharply delineated, started with red to purple discoloration, and then turned to bluish-black, indicating gangrene. Sometimes superficial bullae were formed. With healing the necrotic skin was rejected and a scar was formed. From day 6 to day 10 hematuria was present. There was no palpable mass in the abdomen. On sonography the renal parenchyma was normal, but there was dilatation of the left lower renal pelvis (which persisted, but did not progress later). No urinary infection, no rise in serum creatinine and no fragmentocytes in the peripheral blood were found. Cerebral sonography was normal. Examination of the eyes showed on the right side a central corneal scar, synechisis of the iris, and old hemorrhages in the vitreous. On the left side there was corneal clouding, hyperemia of the iris, and some stripe-like vitreous bleeding. Using routine clotting tests it became evident that each bleeding episode was accompanied by severe hypofibrinogenemia and - less striking - thrombocytopenia (Tab. 1). Fibrin monomers were present, but there was no evidence of severe hyperfibrinolysis.

On day 34 a biopsy from an involved skin area was taken. It revealed extensive hemorrhage into the corium and fibrin thrombi within the capillary blood vessels. There was no vasculitis and no other inflammatory infiltration. Investigation

Table 1. Results of clotting tests.

Date		19.8.	22.8.	23.8.	24.8.	25.8.	26.8.	27.8.	28.8.	29.8.	30.8.	31.8.	
THERAPY	HEPARIN U/24 H		1200	2400	2400	2400	2400	2400	3000	3000	3000		NORMAL RANGE
	FIBRINOGEN MG		300					300			300		
LABORATORY	THROMBOZYTES/µL	53,000	52,000	21,000	122,000	66,000	143,000	83,000	122,000	122,000	51,000	102,000	100,000 - 300,000
	FIBRINOGEN MG/DL	83	40	320	110	140	110	12	250	140	30	50	150-400
	QUICK %	68	48	> 100 ·								80	> 70
	PTT sec	35	42	25								35	< 35
	TZ sec	18	22	21	34	25	39	39	22	38	44	37	< 20
	REPTILASE sec	24	29+)		26	18	25	42	21	30	30	24	< 20
	THROMBINCOGULASE sec		22										< 23
	FIBRINMONOMERS	Ø	++	++	++	++	++		++	(+)	++	++	Ø
	AT III %		70			90							> 70
BLEEDING		+	+++					++++			+++		

+) normalized after addition of fibrinogen

by immune histology revealed deposits of IgG and complement C_3 within arterial vessel walls and fibrin within the lumen of capillary vessels. Serum IgM was below 30 mg/dl and complement levels were within normal limits. A battery of serological studies for infections was negative.

A diagnosis of purpura fulminans and intravascular coagulation was made. Therapy was started on day 6 with escalating doses of heparin from 1200 to 3000 U per day by continuous i.v. infusion. Fibrinogen was substituted by a preparation of Cohn fraction I (300 mg of fibrinogen were given in a volume of 30 ml. As determined later, in this solution the protein C content was 50 % of normal plasma). This therapy was successful inasmuch as bleeding stopped, and fibrinogen and thrombocytes returned to normal levels. It was impossible, however, to prevent further bleedings inspite of continuous heparinization. After the 6th bleeding episode several therapeutic trials were started with aprotinin (Trasylol®), epsilonaminocaproic acid, antithrombin III, fresh frozen plasma, and prednisone. These trials were all unsuccessful and three further relapses occurred.

Use of peripheral veins for i.v. therapy became increasingly difficult. On day 45 oral phenprocoumon was started, beginning with 3 doses of 0.75 mg within 48 hours, and continuing with 2 - 3 doses of 0.75 mg per week according to Quick testing (target level 15 - 30 %). Heparin was stopped on day 49. No further bleeding occurred.

In January 84, when the child was four months old and received 0.75 mg of phenprocoumon on 5 days of the week, the Quick test was 23 %, PTT 42 sec, fibrinogen 165 mg/dl, fibrin monomers negative, antithrombin III 100 %. Blood was drawn for protein C determination. Protein C was measured immunologically with the ELISA-kit of Boehringer Mannheim. In the same specimen the activities of prothrombin (37°) and antithrombin III (25°) were measured using chromogenic substrates (test combinations of Boehringer Mannheim). Factor VII activity was also determined with a photometric assay (unpub-

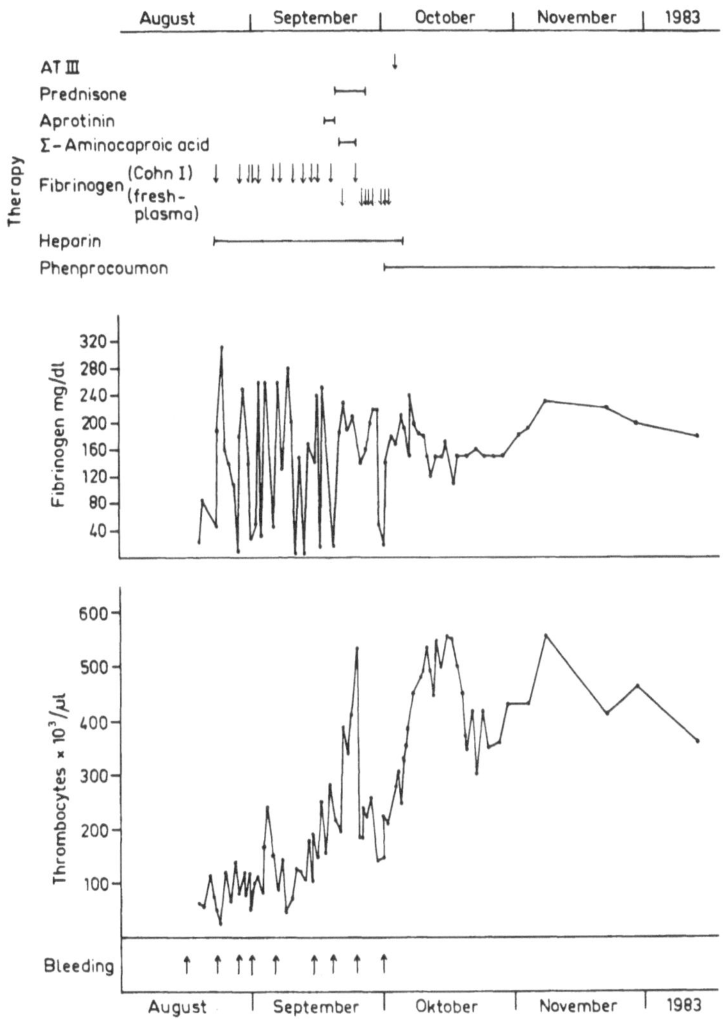

Fig. 2. Clinical course of the patient in relation to therapy. Each bleeding is accompanied by severe hypofibrinogenemia and moderate thrombocytopenia. Cure after initiation of phenprocoumon.

lished). The results are given in Table 2. A stored plasma sample was available from the 47th day of life, the third day of phenprocoumon treatment. Protein C concentration was less than 1 % in this specimen.

Table 2. Protein C, antithrombin III, and vitamin K-dependent coagulation factor levels in the patient, his parents, and in a control infant.

	Protein C	Factor VII	Prothrombin	Antithrombin III
Father V.S. born 03-11-61	36 %	100 %	20 U/ml	11 U/ml
Mother R.S. born 01-22-63	37 %	67 %	17 U/ml	11 U/ml
Patient C.S. born 08-17-83	1 %	20 %	4.7 U/ml	10 U/ml
Control patient of same age	46 %	49 %	12 U/ml	10 U/ml

We have seen the child again in September 1984. He was still receiving 0.75 mg of phenprocoumon on 5 days of the week. Quick test was 21 %, PTT 45 sec, fibrinogen 240 mg/dl, fibrinogen monomers absent, reptilase time 17 sec, factor VIII 260 %, factor V 880 % (!), protein C 1 %, SGOT 17, SGPT 25 U/l. Developmental age was 9 - 13 months depending on the function tested. Cranial CT and EEG were normal. He was blind on his right eye, and vision was extremely reduced on the left side.

Discussion

Purpura fulminans may occur at different ages and the pathogenesis has been related to the Shwartzman phenomenon (1, 2). Whereas several older children were cured with heparin (4, 5), all newborn babies with this disorder have died (1, 2, 3). Protein C deficiency as a possible cause was first described by Branson et al. (3). In the meantime more newborn babies with purpura fulminans and protein C deficiency have been diagnosed (6). On the other hand, Seligsohn and co-workers reported on a protein C deficient infant presenting with massive venous thrombosis without purpura (7). Thus the clinical picture of severe congenital protein C deficiency does not seem to be uniform. Nevertheless it can be assumed that in purpura fulminans of the newborn protein C deficiency is the cardinal pathogenetic factor.

The diagnosis in our patient was made when he was already receiving phenprocoumon, and we did not dare interrupt this treatment for diagnostic purposes. Nevertheless we believe that the depression of protein C in our patient is in excess of that expected from vitamin K antagonism of phenprocoumon. Our data are consistent with homozygosity of the child and heterozygosity of both the parents. The disease of the child began in utero.

Phenprocoumon was started because of technical difficulties with the venous route and not because of pathophysiological considerations. Phenprocoumon has not altered the course of a baby reported previously (1). In fact, coumarin anticoagulants may induce skin necrosis in patients with (heterozygous) protein C deficiency (8). It is believed that in these heterozygous patients further protein C depression by coumarin antecedes drug-induced prothrombin deficiency, thus aggravating transiently the imbalance between anticoagulant and procoagulant factors. This mechanism, if it is correct, may be irrelevant for patients such as ours who are suffering from maximal protein C deficiency from the beginning. It is

striking to see how abruptly phenprocoumon stopped all bleeding episodes in our patient (Fig. 2) as well as in Branson's (3). We have continued this treatment for one year now, and the patient was in continuous remission all the time.
The question remains why heparin was unsuccessful as in Branson's patient (3). Recently it was shown that heparin enhances the inactivation of protein C in normal plasma (9). It is doubtful, however, if such an effect can play a role in a patient who is depleted of protein C anyway.

References

1. Trüeb, O., Willi, H., Siebenmann, R., Hitzig, W.H. Frick, P.G.: Helv. paediat. Acta 19, 223-242 (1964)

2. Van der Horst, R.L.: Arch. Dis. Childh. 37, 436-441 (1962)

3. Branson, H.E., Katz, J., Marble, R. Griffin, J.H.: Lancet II, 1165-1168 (1983)

4. De Koning, J., Frederiks, E., Kerkhoven, P.: Helv. paediat. Acta 27, 177-185 (1972)

5. Naveh, Y., Takarsky, I., Friedman, A.: Helv. paediat. Acta 27, 193-196 (1972)

6. Griffin, J.H.: personal communication (1984)

7. Seligsohn, U., Berger, A., Arend, M., Rubin, L., Attias, D., Zivelin, A., Rapaport, S.I.: New Engl. J. Med. 310, 559-562 (1984)

8. Broekmans, A.W., Bertina, R.M., Loeliger, E.A., Hofman, V., Klingemann, H.-G.: Thromb. Haemostas. 49, 251 (1983)

9. Comp, P.C., Nixon, R.R., Esmon, C.T.: Blood 63, 15-21 (1984)

PROTEIN C IN PATIENTS UNDER L-ASPARAGINASE TREATMENT

Theodor H. Schöndorf and Irene Witt

Zentrum für Innere Medizin der Justus-Liebig-Universität
D-6300 Giessen
Biochemisches Labor der Universitäts-Kinderklinik
D-7800 Freiburg

Introduction

L-Asparaginase is used in the treatment of acute lymphoblastic leukemia and some malignant lymphomas. The enzyme L-asparaginase inhibits the growth of cells by converting L-asparagine to L-aspartic acid and ammonia. The depletion of asparagine accounts for the antineoplastic effect. The proliferant cells of the bone marrow and of the intestinal epithelium are less affected than the lymphoblasts.

The treatment with L-asparaginase which is derived from E. coli can be accompanied by various side-effects such as fever, anaphylactic reactions, pancreatitis and liver damage. In addition, haemorrhages due to hypofibrinogenemia were observed previously as a complication during the L-asparaginase treatment (1).

In earlier clinical studies intravascular coagulation has also been reported. However, since L-asparaginase is restricted to the treatment of acute lymphoblastic leukemia and is not used for other acute leukemias such as promyelocytic leukemia, a consumption coagulopathy cannot be found.

Studies with labelled fibrinogen demonstrated that its decrease is induced by an inhibited synthesis (2). The inhibition of protein synthesis by L-asparaginase causes not only a deficiency of coagulation factors such as fibrinogen,

Protein C - Biochemical and Medical Aspects

factor IX and factor XI (3, 4) but also the major plasma inhibitor antithrombin III and the main protein of the fibrinolytic system - plasminogen - are regularly reduced (5, 6, 7).
In recent years clinical observations of thrombotic events associated with L-asparaginase treatment have been more frequently reported. Priest et al. (8, 9) reviewed thrombotic complications in children receiving chemotherapy including L-asparaginase. 16 out of 1400 children were observed to have thrombotic phenomena; this is 1.4 % complications. Conard et al. (10) reported 10 % thrombotic events in 40 adults. Our own study showed an incidence of 4/15 patients (27 %) with venous or arterial thrombotic complications (11).
Low levels of antithrombin III induced by an impaired synthesis were considered as an important pathogenetical factor of the thrombotic events under or after cessation of L-asparaginase treatment.
The importance of a congenital protein C deficiency is now recognized as an important factor of inherited thrombophilia (12, 13). The aim of our study was to investigate whether and to what extent the protein C concentration is decreased by L-asparaginase treatment and whether a continuous or intermittent administration of L-asparaginase influences the protein C concentration in the same way.

Methods

Protein C antigen concentration was measured by an Elisa kit according to the recommendations of the manufacturer (Boehringer Mannheim, FRG).
Activities of antithrombin III and prothrombin were determined by chromogenic substrate assays (Boehringer Mannheim, FRG).
Fibrinogen was measured functionally with the clotting test according to Clauss and α_2-macroglobulin immunologically with the Immunchemical System (Beckman Instruments, Munich, FRG).

Patients

Tab. 1 summarizes the clinical information of the patients. Seven adult patients - 5 males and 2 females - were studied. Their mean age was 32 years. The patients had no previous thromboembolic complications. 4 patients suffered from acute lymphatic leukemia, 2 patients from acute undifferentiated leukemia and 1 patient from a lymphoblastic lymphoma. 8,000 - 10,000 U L-asparaginase from E. coli were administered per day by continuous infusion usually during 10 - 12 consecutive days.

The 3 pediatric patients suffered from acute lymphatic leukemia. 8 x 10,000 U L-asparaginase per m^2 were infused intermittently every 3rd day for 3 hours.

Table 1. Clinical data of the investigated patients.

PATIENTS		AGE (years)	HEMATOLOGICAL DISEASE	L-ASPARAGINASE
Internal Medicine				
G. H.	♀	16	Acute lymphatic leukemia (ALL)	8 000 - 10 000 U / day as infusion / 6 h over 10 - 12 days
M. D.	♂	22		
H. B.	♂	23		
W. J.	♂	18		
L. Sch.	♀	49	Acute undiff. leukemia	
W. B.	♂	46		
W. D.	♂	50	Lymphoblastic lymphoma	
Pediatrics				
S. V.	♂	3	Acute lymphatic leukemia (ALL)	8 x 10 000 U / m^2 as infusion / 3 h every 3rd day
D. B.	♀	5		
D. M.	♂	15		

Results

The fibrinogen concentration and the antithrombin III activity decreased progressively and significantly in the adult patients in the initial phase of L-asparaginase therapy, as is known from previous investigations. The decline corresponds to the half-life time of the proteins. The concentration of protein C antigen also decreased markedly in all patients investigated at the beginning of the therapy.

The behaviour of protein C is demonstrated in the following examples. The data of the patient shown in Fig. 1 indicate that the lowest protein C concentration of 40 % was reached on the 6th day of the continuous L-asparaginase treatment. Thus the protein C decrease does not correspond to its half-life time of a few hours. It has to be assumed that the synthesis is reduced but not completely interrupted.

Antithrombin III activity was lowered to a critical value of 5.8 U/ml plasma on the 5th day, therefore 1000 U antithrombin III were substituted as concentrate. 2 units of fresh frozen plasma were transfused twice, however, this was not followed by a relevant increase of the plasma protein C concentration.

The next example (Fig. 2) shows again the decrease of the antithrombin III activity. Antithrombin III was below 6 U/ml plasma on the 4th day of the asparaginase treatment. An increase of the plasma antithrombin III was obtained by an antithrombin III substitution.

A pronounced decrease of protein C, which was reduced below 25 % of normal, was found on the 5th day of treatment. The pre-treatment level was 70 %. In addition to the reduction of protein C the vitamin K-dependent prothrombin activity decreased concomitantly, but the decline was less pronounced.

The third example (Fig. 3) reveals again a significant reduction of the two coagulation inhibitors antithrombin III and protein C. The lowest levels were observed on the 5th day of the continuous L-asparaginase treatment. As antithrombin III dropped below 4 U/ml plasma, antithrombin III concentrates

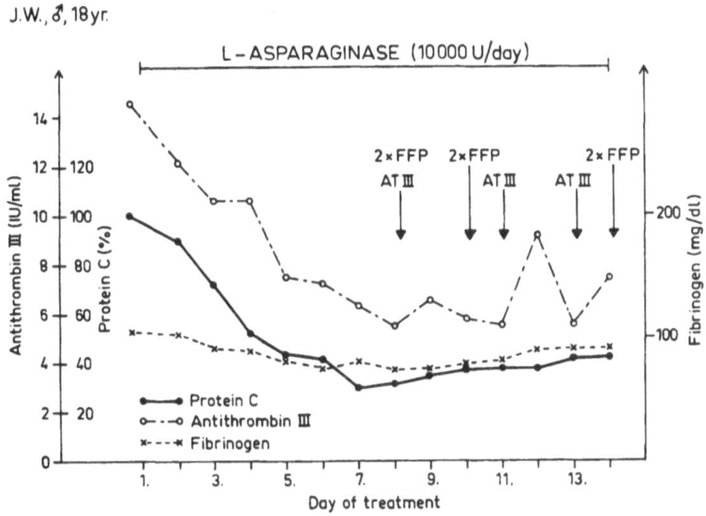

Fig. 1. Protein C antigen concentration, antithrombin III activity and fibrinogen concentration in an adult patient under continuous L-asparaginase treatment. Substitution with antithrombin III concentrate and fresh frozen plasma.

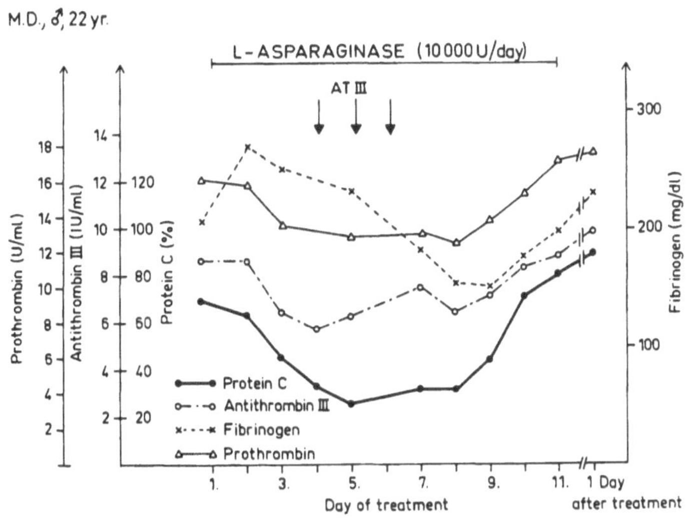

Fig. 2. Protein C antigen and fibrinogen concentrations together with antithrombin III and prothrombin activities in an adult patient under continuous L-asparaginase treatment. Substitution with antithrombin III concentrate.

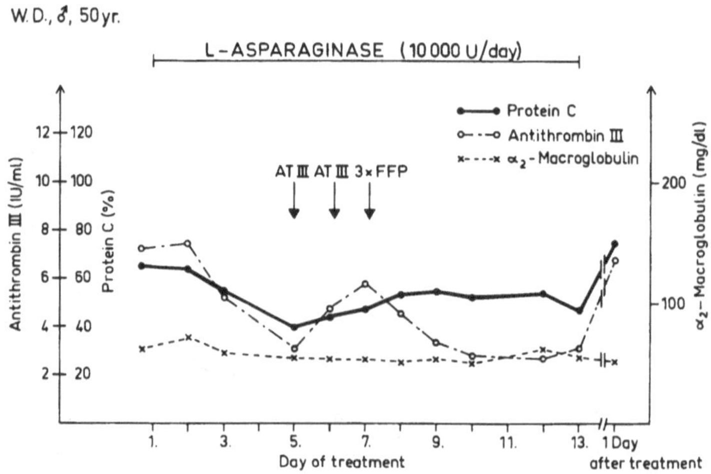

Fig. 3. Protein C antigen and α_2-macroglobulin concentrations as well as antithrombin III activity in a patient under continuous L-asparaginase treatment. Substitution with antithrombin III concentrate and fresh frozen plasma.

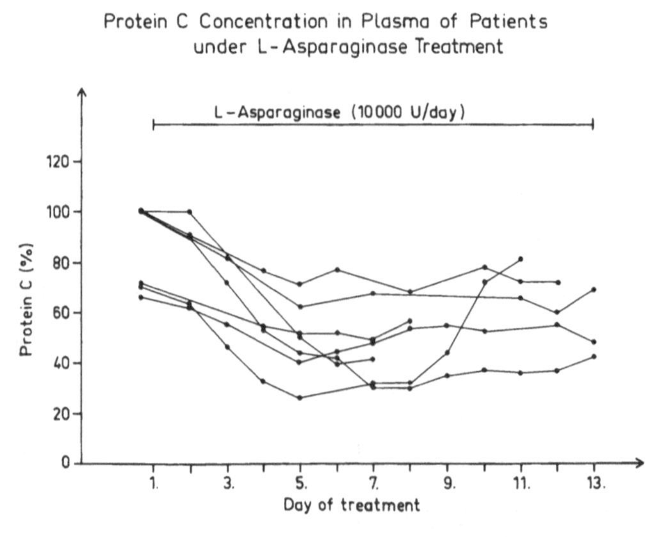

Fig. 4. Behaviour of protein C antigen in 7 adult patients under continuous L-asparaginase treatment.

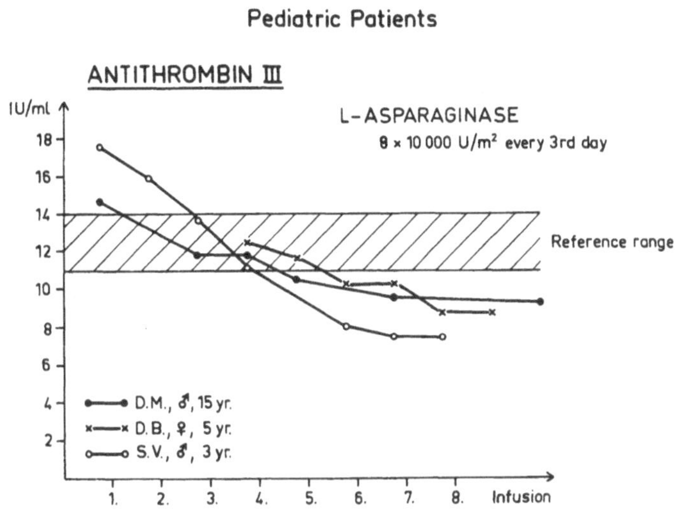

Fig. 5. Antithrombin III activity in 3 pediatric patients under intermittent L-asparaginase treatment.

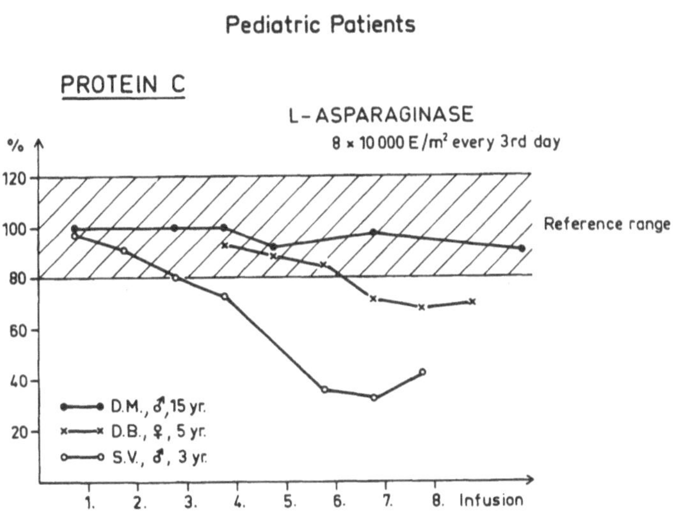

Fig. 6. Protein C antigen concentration in 3 pediatric patients under intermittent L-asparaginase treatment.

were given. An elevation of the protein C concentration to subnormal levels was obtained after the transfusion of 3 units fresh frozen plasma.

The inhibition of the protein synthesis by L-asparaginase is restricted to the liver because the α_2-macroglobulin concentration remained stable during the 13 days of treatment. α_2-macroglobulin is assumed to be synthesized in macrophages and not in the liver. This behaviour of α_2-macroglobulin was also confirmed in the other patients.

Fig. 4 summarizes the protein C antigen concentrations of our 7 adult patients during L-asparaginase therapy. Under the continuous administration of L-asparaginase, the lowest levels of protein C were most often observed between the 5th and 7th day. In the following days no further reduction occurred. Protein C usually decreased by 30 % from variable pretreatment levels.

Increase of protein C concentration which could be seen during the later L-asparaginase treatment is due to the substitution therapy with fresh frozen plasma.

None of the 7 patients developed a clinically overt thromboembolic event. This observation might be explained by the fact that none of the patients had predisposing risk factors for thromboembolic complications.

Thromboembolic complications occur very rarely during L-asparaginase treatment in children as reviewed by Priest (9), although antithrombin III decreases pronouncedly during administration of the drug.

Fig. 5 shows the antithrombin III activity in the plasma of 2 children and 1 adolescent. The most pronounced decrease of antithrombin III activity occurred in the 3 year old child. Pretreatment antithrombin III activities were elevated above the reference range due to the previous and concomitant cortisol therapy.

Also the protein C concentration decreases under intermittent L-asparaginase treatment and cortisol therapy in children (Fig. 6). Besides the pronounced antithrombin III reduction the protein C concentration diminished significantly in the 3 year old child.

Conclusions

Our study demonstrates that the synthesis of the coagulation inhibitor protein C is also impaired during treatment with L-asparaginase. The protein C concentration decreased considerably during the long-term administration of the drug. Adult patients revealed the lowest protein C concentrations between the 5th and 7th day of continuous L-asparaginase treatment. An intermittent dose schedule employed in pediatric patients can also reduce the protein C concentration. In addition to the already known low concentration and activity of antithrombin III, this means a pronounced reduction of the hemostatic inhibitor capacity.

Clinical signs of thromboembolic events were not observed in our patients although both coagulation inhibitors were decreased to values which are associated in familiar deficiencies with an increased risk of thromboemboli.

Presumably the reason is that none of our patients had an additional clinical risk for thromboembolic complications. Moreover under L-asparaginase fibrinogen and further procoagulatory proteins are also synthesized less. Depending on the pretreatment activities of the procoagulatory factors and inhibitors, which can also be influenced by other medication, e.g. cortisol, and the variable degree of synthesis inhibition, the balance of the hemostatic system can be shifted either in the direction of procoagulation with a tendency to thrombosis or in the direction of anticoagulation with a tendency of hemorrhage. At the end of L-asparaginase treatment, a decrease of the inhibitors prevails more frequently, thus promoting a thrombotic tendency. Together with clinical risk factors thromboembolic complications may occur.

Based on our data, a general recommendation for replacement therapy cannot be given. A substitution for antithrombin III or protein C in adults receiving L-asparaginase seems to be indicated if the decrease of the inhibitors below normal values is demonstrated and if predisposing factors for the development of thromboemboli co-exist. Risk factors such as advanced age, history of previous thromboemboli, immobilization, obesity or low levels of antithrombin III and/or protein C prior to L-asparaginase treatment have to be considered for the individual indication of replacement therapy with fresh frozen plasma or antithrombin III concentrate.

References

1. Bettigole, R.E., Himelstein, E.S., Oettgen, H.F., Clifford, G.O.: Clin. Res. 17, 399 (1969).
2. Bettigole, R.E., Himelstein, E.S., Oettgen, H.F., Clifford, G.O.: Blood 35, 195-200 (1970).
3. Gralnick, H.R., Henderson, E.: Cancer 27, 1312-1320 (1971).
4. Lechler, E., Schumacher, K., Hirschmann, W.D.: Med. Welt 22, 127-133 (1971).
5. Liebman, H.A., Wada, J.K., Patch, M.J., McGehee, W.: Cancer 50, 452-456 (1982).
6. Pitney, W.R., Phadke, K.P., Dean, S.: Lancet i, 493-494 (1980).
7. Vellenga, E., Mulder, N.H., Nieweg, H.O.: Lancet i, 649-650 (1980).
8. Priest, J., Edson, J.R., Levitt, C., Tubergen, D., Steinherz, P., Cairo, M., Bishop, A., Sitarz, A., White, L., Coccia, P., Krivit, W.: ASCO Abstracts, C-612 (1981).
9. Priest, J.R., Ramsay, N.K.C., Bennett, A.J., Krivit, W., Edson, J.R.: J. Pediatr. 100, 990-995 (1982).
10. Conard, J., Cazenave, B., Maury, J., Horellou, M.H., Samama, M.: Lancet i, 1091 (1980).
11. Schöndorf, T.H., Pralle, H., Graubner, M., Lasch, H.G.: Thromb. Haemostas. 50, 50 (1983).
12. Griffin, J.H., Evatt, B., Zimmerman, T.S., Kleiss, A.J., Widemann, C.: J. Clin. Invest. 68, 1370-1373 (1981).
13. Bertina, R.M., Broekmans, A.W., van der Linden, I.K., Mertens, K.: Thromb. Haemostas. 48, 1-5 (1982).

PROTEIN C IN VARIOUS LIVER DISEASES

Gerd Oehler, Karl Matthes
Zentrum für Innere Medizin der Justus Liebig-Universität,
D-6300 Giessen

Introduction

Protein C is assumed to be synthesized by the liver. In one study it has been shown that protein C levels are decreased especially in patients with severe liver diseases (1). At present there are not so many data about protein C values in different liver diseases (2). Therefore we classified the liver diseases into numerous subgroups in order to reveal further information about the implications of an impaired liver function on protein C. The protein C values were compared with cholinesterase and prealbumin which indicate the capacity of the liver to synthesize proteins.

Methods

69 patients with various acute and chronic liver diseases were examined. Acute hepatitis was diagnosed on the basis of clinical and chemical criteria. All other diseases were proven histologically and by laparoscopy. The following diagnoses were established:

acute (viral) hepatitis	7 patients
acute hepatitis after recovery	7 patients
chronic active hepatitis	11 patients

Protein C - Biochemical and Medical Aspects

chronic persistent hepatitis	14 patients
chronic active hepatitis in quiescent state at the time of examination	6 patients
alcoholic hepatitis	7 patients
alcoholic hepatitis in quiescent state	7 patients
liver cirrhosis	13 patients
cirrhosis in quiescent state	4 patients
hemochromatosis	6 patients
biliary cirrhosis and chronic active hepatitis with cholestasis	6 patients

(This group was characterized by positive antimitochondrial antibodies.)

In many patients we have carried out several measurements on subsequent days.
Protein C was determined in venous citrated plasma (1:10) with ELISA-Test (3) of Boehringer Mannheim according to the instructions of the producer. Cholinesterase was measured with an automated enzyme assay. Prealbumin was determined by radial immunodiffusion.

Results

In acute hepatitis protein C was reduced to about 50 % of normal. After recovery when transaminases had returned to normal, protein C values were 103 %. Patients with chronic active hepatitis had protein C levels of 76 ± 15 % whereas in chronic persistent hepatitis the mean protein C level was 90 ± 10 %. Similar values were found in chronic active hepatitis in the quiescent phase (Fig. 1).
In alcoholic hepatitis protein C concentrations were 69 ± 32 % and in the quiescent phase a normal protein C range of about 100 % was found (Fig. 2).

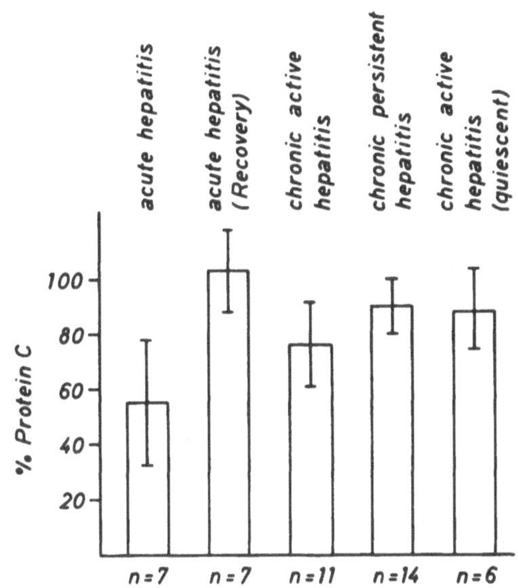

Fig. 1. Protein C in liver diseases

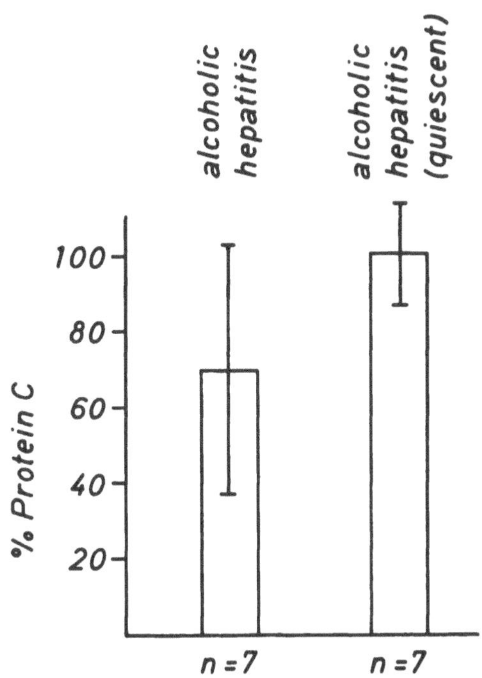

Fig. 2. Protein C in liver diseases

In liver cirrhosis the lowest protein C values were measured. The concentrations were 30 %. When a quiescent state of cirrhosis was reached protein C concentrations were 94 %. Patients with hemochromatosis, all treated by phlebotomies, had protein C levels of 77 ± 23 %. In patients with biliary cirrhosis and chronic active hepatitis with cholestasis with positive antimitochondrial antibodies, the protein C values were relatively high (92 ± 13 %) (Fig. 3).

The protein C levels were not significantly related to cholinesterase and prealbumin.

As already mentioned surprisingly high protein C values were measured in patients with biliary cirrhosis. Therefore we separated this group. We divided the remaining patients into 3 subgroups with different cholinesterase values (Fig. 4). We found significant differences of protein C in the 3 groups. Although the cholinesterase activities of the biliary disease group was in an intermediate range protein C was very high.

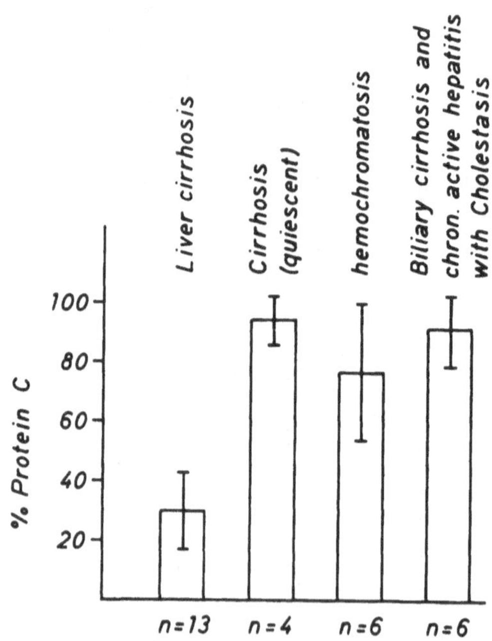

Fig. 3 Protein C in liver diseases

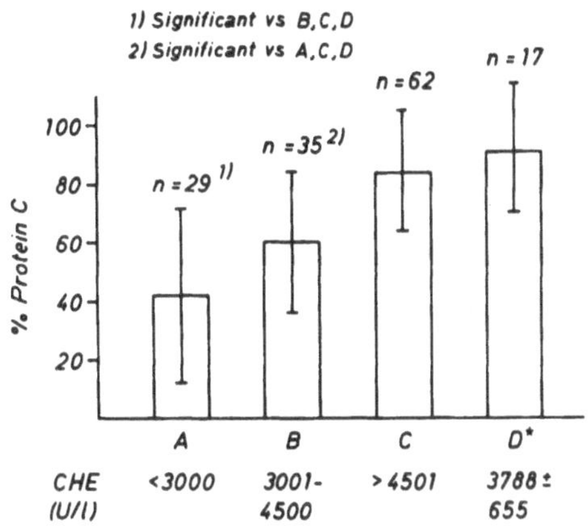

Fig. 4. Protein C in liver diseases. Subgroups on the basis of ChE. *D = biliary cirrhosis and chronic active hepatitis with cholestasis.

Discussion

Our investigations have shown that some liver diseases are accompanied by low protein C levels. According to Mannucci's results (1), marked decrease of protein C was found in liver cirrhosis. In our patients even lower values were measured (30 % vs. 50 % in Mannucci's study).
Also in acute hepatitis a substantial decrease of protein C was found while in the other diseases such as chronic active hepatitis, alcoholic hepatitis and hemochromatosis only minimal decreases were observed. It is of interest that in acute hepatitis and in alcoholic hepatitis protein C returns to normal after recovery from the acute disorder. There was also a (insignificant) difference between chronic active hepatitis in the florid phase and the quiescent phase.

No significant correlation could be demonstrated between protein C and cholinesterase or prealbumin which are examplary for proteins synthesized by the liver. This points to additional influences on protein C concentration. Probably in some cases a disseminated intravascular coagulation brings about an increased turnover of protein C.
For clinical conditions protein C values can be assigned to different groups of liver diseases characterized by the efficiency to synthesize proteins like cholinesterase. In 3 subgroups which we formed the concentrations of protein C were significantly different.
It is noteworthy that patients with biliary cirrhosis had very high protein C levels. We cannot give any evident explanation of this fact. It should be mentioned that AT III - the other important inhibitor of the coagulation system - is also increased in cholestasis.

References

1. Mannucci, P.M., Vigano, S.: Lancet II, 463-466 (1982).

2. Pabinger-Fasching, I.: Hämostaseologie 4, 50-53 (1984).

3. Amiral, J.: Symposium: Recent developments on protein C, Paris, October 1983

PROTEIN C DETERMINATION IN PATIENTS WITH THROMBOTIC DISEASE AND DISSEMINATED INTRAVASCULAR COAGULATION

Ingrid Pabinger-Fasching, Ulrike Stoffels, Klaus Lechner
I. Medizinische Universitäts-Klinik
A-1090 Wien

We want to report on protein C levels in three different groups of patients:

I Patients with a history of venous thrombosis and/or pulmonary embolism

II Patients with nephrotic syndrome

III Patients with acute myeloblastic leukemia with and without disseminated intravascular coagulation

For protein C determination we used the Laurell rocket immunelectrophoresis (1, 2). The protein C antibody was a kind gift of Dr. Bertina (Leiden, The Netherlands). Protein C activity was determined by the method described by Francis et al. 1984 (3), which was slightly modified: The vitamin K-dependent factors were adsorbed by $Al(OH)_3$ and eluted by Na-K-phosphate buffer and activated by alpha-thrombin. Activation was stopped by AT III and heparin, heparin was neutralized with protamine and protein C activity was determined by its ability to prolong the aPTT.

I. Protein C in patients with a history of venous thrombosis and/or pulmonary embolism

I.1 Prevalence

We investigated 338 patients with at least one major thrombo-

Table 1. Prevalence of hereditary protein C and AT III deficiency in patients with a history of venous thrombosis

	Protein C	AT III
Investigated patients	338	752
DEFINITE hereditary deficiency*	5(1.5%)	10(1.33%)
PROBABLE hereditary deficiency**	0	9(1.2%)

* 2 or more deficient family members

** deficiency proven and thromboembolic events in the family history

embolic event and found 5 patients with proven hereditary protein C deficiency; that's a prevalence of 1.5 % (Tab. 1). Out of 752 patients with a history of venous thromboembolism, in whom protein C was determined, we found 10 with definite and 9 with probable AT III deficiency. The prevalence of hereditary protein C deficiency seems to be equal or slightly less than AT III deficiency.

Subsequent investigations of the family members of the 5 protein C deficient patients revealed further 15 patients with protein C deficiency.

I.2 <u>Clinical picture of hereditary protein C deficiency</u>

11 out of 20 protein C deficient patients experienced at least one thromboembolic event, most of these patients showed severe thrombotic tendency and their medical history was typical for protein C deficiency as it is described in literature.

I only want to mention two families with special manifestations of this disease:

- One male patient experienced mesenteric vein thrombosis (MVT). The father of this patient had died of MVT, diagnosis was established by autopsy. We know that MVT occurs in 8 % of thromboembolic events in patients with AT III deficiency and thromboembolism (4). There seems to be no main difference in the site of thrombosis between protein C deficiency and AT III deficiency.
- The proposita of an Egyptian protein C deficient family developed deep venous thrombosis (DVT) after delivery. After streptokinase therapy and surgical thrombectomy, oral anticoagulant therapy was instituted on the 6th postoperative day with 12 mg of phenprocoumon. On the fourth day of coumarin treatment she developed painful discoloration of three toes and on the 5th day the full picture of coumarin necrosis was present. The family history revealed no thrombotic tendency, although two sisters (29 and 26 years of age) were protein C deficient. One had had a normal pregnancy and delivery. Protein C in the mother was normal, the father had died after leg amputation, thromboembolic episodes were not recalled.

I.3 Therapeutic aspects

From the theoretical point of view, heparin should be an effective anticoagulant. Our experience in protein C deficient patients support this assumption:

4 patients with hereditary protein C deficiency underwent surgery (appendectomy in adulthood, hip surgery, knee surgery, breast surgery) under conventional low dose heparin prophylaxis (3 x 5000 IU Dep. Heparin s.c.) and none of the patients developed any thrombotic complications. One patient with MVT recovered quickly under heparin treatment.

We confirm that oral anticoagulant treatment is effective in preventing major thromboembolic events. 4 of our protein C deficient patients were observed under long-term oral anticoagulant treatment, and none of these patients experienced a major thromboembolic event. In one patient we observed two episodes of superficial thrombophlebitis despite stable anticoagulant treatment.

II Protein C levels in patients with nephrotic syndrome

Patients with nephrotic syndrome are known to have a greatly increased risk of thrombosis. Acquired AT III deficiency due to urinary loss is known as the important cause for the thrombophilic state (5, 6). Since protein C has a low molecular weight similar to albumin and AT III, it may be lost in the urine of patients with severe proteinuria and consequently its plasma level may decrease. We therefore examined the plasma and the urine protein C levels of 8 patients with nephrotic syndrome (proteinuria more than 5 g/day). The study included 3 children and 5 adults, the median age was 30 years, the range was 4 - 75 years. At the time of examination one patient had experienced deep vein thrombosis and another patient was on treatment with corticosteroids.
The patient's plasma protein C:Ag concentration was significantly elevated compared to 13 healthy normal adults ($p < 0.001$) (Fig. 1). In 6 of 8 patients the levels were above the normal range. Protein C activity determinations were performed in three patients and revealed elevated protein C activity well comparable to protein C:Ag concentrations. AT III:Ag levels were, as expected, decreased. There was a significant inverse correlation of protein C:Ag to AT III:Ag plasma levels ($r = -0.78$, $p < 0.05$) (Fig. 2). Plasma F II:Ag concentration was elevated in five patients, but a significant correlation with protein C:Ag could not be demonstrated.

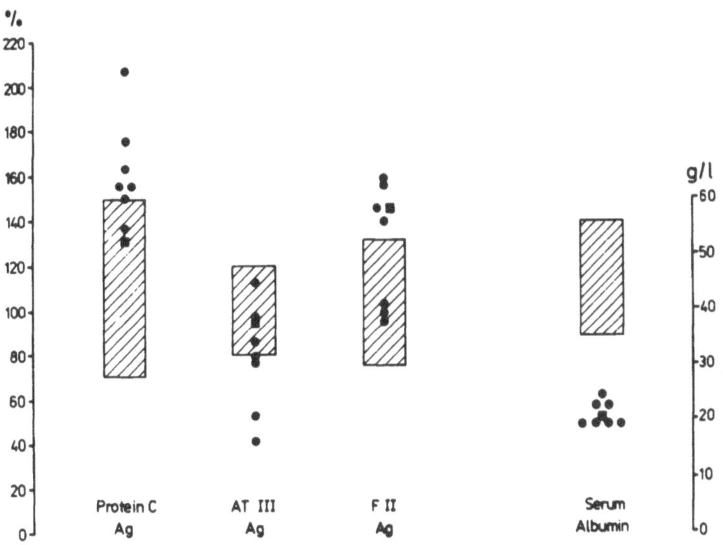

Fig. 1. Protein C:Ag, AT III:Ag, F II:Ag and serum albumin in 8 patients with nephrotic syndrome.

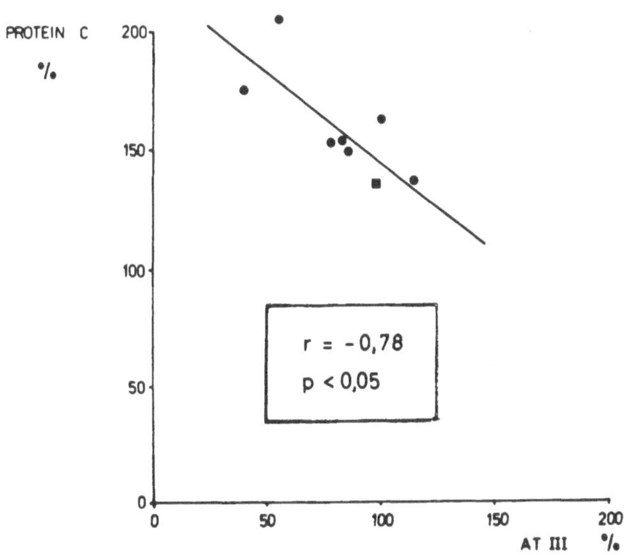

Fig. 2. Correlation of protein C to AT III in 8 patients with nephrotic syndrome.

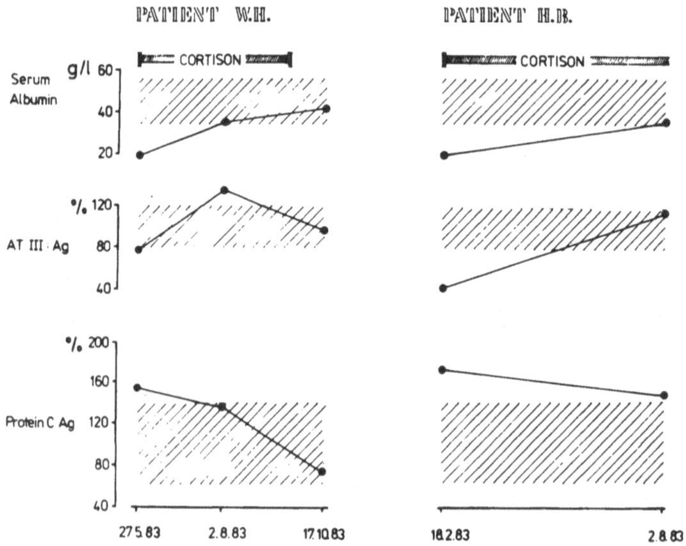

Fig. 3. Follow-up studies of serum albumin, AT III:Ag and protein C:Ag in two children.

Urine samples of two children were repeatedly analysed for protein C:Ag. No protein C:Ag could be detected in six different urine samples of two patients, whereas AT III:Ag was repeatedly detectable.

Follow-up studies were performed in two children under steroid treatment, which lead to a complete remission in one patient and to a partial remission in the second child. Serum albumin and AT III:Ag rose to normal levels, whereas protein C:Ag fell from elevated pretreatment levels towards normal.

We conclude from our data that changes in protein C plasma levels cannot be reponsible for the high thrombotic tendency in patients with nephrotic syndrome.

III Protein C in patients with acute myeloblastic leukemia with and without disseminated intravascular coagulation

Protein C has been reported to be reduced in patients with disseminated intravascular coagulation (DIC), which occurs (7, 8) frequently in patients with acute myeloblastic leukemia (AML). We investigated 36 patients with AML, 24 without DIC and 12 with DIC.

Diagnosis of DIC was established when:

1. fibrinogen was lower than 150 mg/dl,
2. FPA was elevated,
3. the ethanol gelation test was positive and/or FDP were increased.

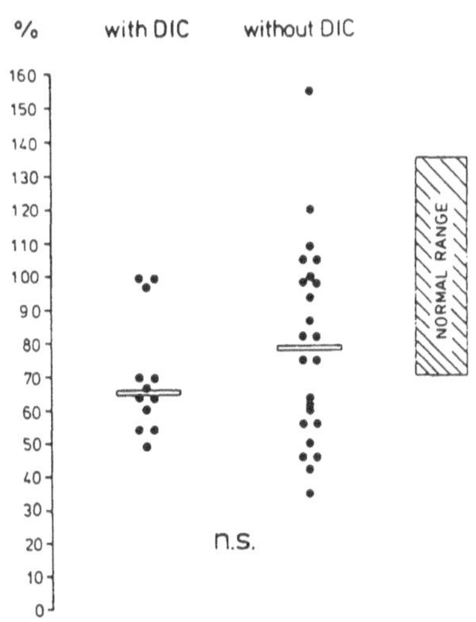

Fig. 4. Protein C in patients with AML.

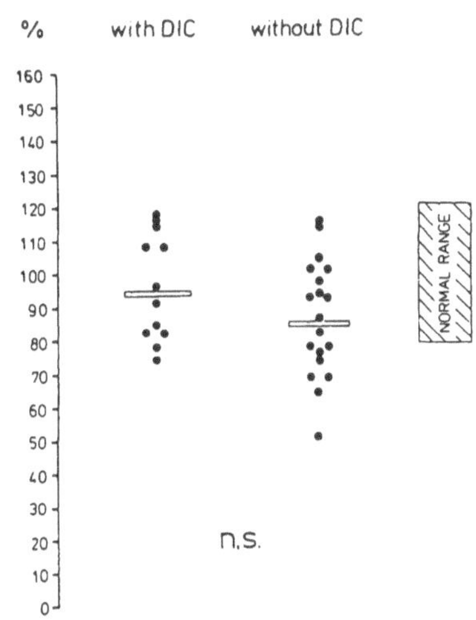

Fig. 5. AT III in patients with AML.

Median protein C:Ag concentration in patients with AML with and without DIC was significantly lower compared to healthy adults. In patients with DIC median protein C level was 66 % (range 48 - 100 %) and was lower than in patients without DIC (median 78.5 %, range 35 - 155 %) (Fig. 4). The range of protein C concentrations was wide in both groups, the difference was not significant.

AT III activity was almost the same in patients with and without DIC (Fig. 5).

A highly significant correlation of AT III and protein C was observed, irrespective of whether DIC was present or not (Fig. 6).

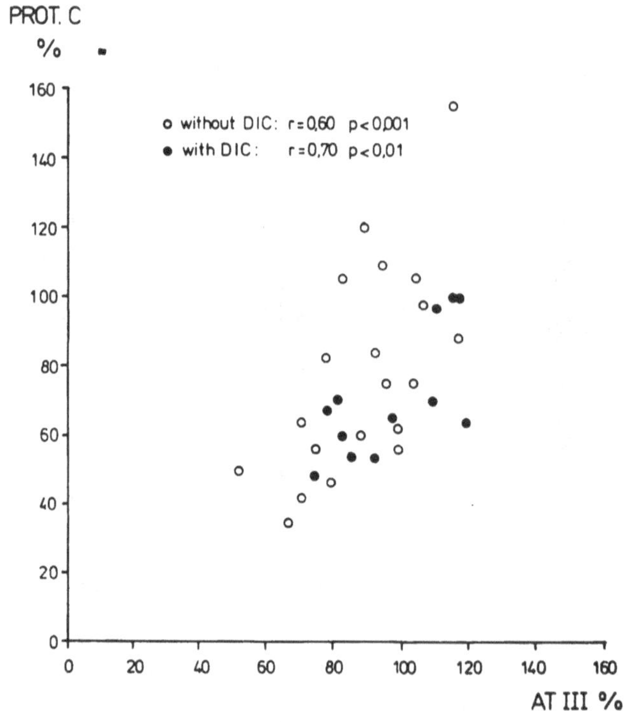

Fig. 6 Correlation of protein C with AT III in AML.

There was a significant correlation betwen protein C and Normotest in patients without DIC, but no correlation in those with DIC (Fig. 7). F II:Ag did not show any correlation with protein C.
Therefore, it is likely that in patients without DIC the plasma concentration of protein C and AT III reflects the protein synthesis capacity of the liver.
In patients with DIC probably other factors than liver function may influence plasma protein C concentration. These factors could not be identified since infection and bleeding were not related to the protein C level. From our data it does not appear that determination of protein C in patients with AML is of great diagnostic value.

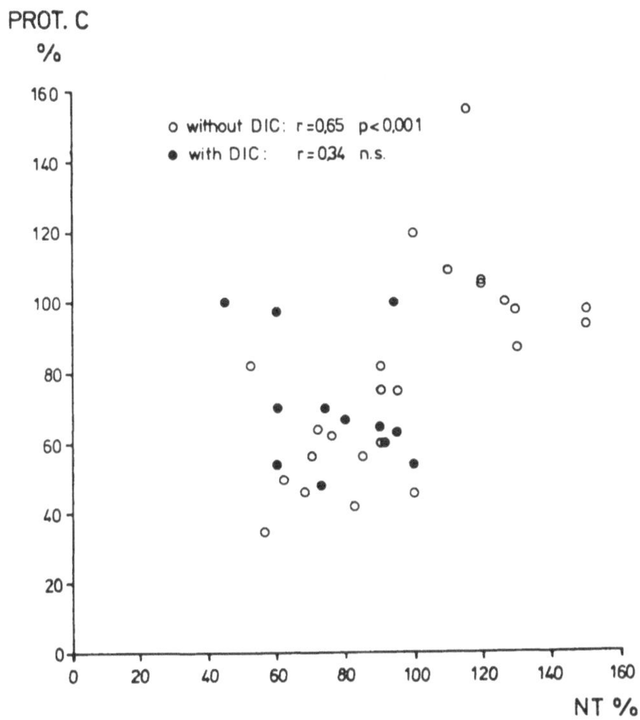

Fig. 7. Correlation of protein C with Normotest in AML.

References

1. Bertina, R.M., Broekmans, A.W., van der Linden, I.K., Mertens, K.: Thromb. Haemostas. 48, 1-5 (1982).

2. Pabinger-Fasching, I., Bertina, R.M., Lechner, K., Niessner, H., Korninger, Ch.: Thromb. Haemostas. 50, 810-813 (1983).

3. Francis, B.R., Patch, M.J.: Thromb. Res. 32, 605-613 (1983).

4. Thaler, E., Lechner, K.: Clin. Haematol. 10, 369-390 (1981).

5. Kaufmann, H.R., Veltkamp, J.J., van Tilburg, H.N., van Es, L.A.: Amer. J. Med. 65, 607-613 (1978).

6. Thaler, E., Balzar, E., Kopsa, H., Pinggera, W.F.: Hemostasis 7, 257-272 (1978).

7. Mannucci, P.M., Vigano, S.: Lancet 2, 463-467 (1982).

8. Griffin, J.H., Mosher, D.F., Zimmerman, T.S., Kleiss, A.J.: Blood 60, 261-264 (1982).

INVESTIGATIONS OF THE INCIDENCE OF DIFFERENT INHERITED FIBRINOLYSIS DISORDERS IN JUVENILE THROMBOSIS PATIENTS

Inge Scharrer

Abteilung für Angiologie, Zentrum der Inneren Medizin,
Klinikum der Universität, Frankfurt/Main, FRG

In patients with recurrent thrombosis up to 45 years and with family disposition, very often only 1 or no risk factor can be found.

Therefore hereditary defects of coagulation and fibrinolysis system presumably play a key role in thrombogenesis. Hemostaseological factors which are involved in the occurrence of recurrent venous thrombosis are antithrombin III deficiency and abnormal AT III molecule, plasminogen deficiency and abnormal plasminogen molecule, decreased release and content of vascular plasminogen activator or increased TPA-inhibitor, dysfibrinogenemia, protein C deficiency and abnormal protein C molecule. The thrombosis presents striking clinical characteristics: recurrence, high incidence of pulmonary embolism, therapy resistance to lysis, familiar predisposition and first manifestation at the age of $<$40 years.

Tab. 1 shows the incidence of inherited fibrinolysis disorders in 200 examined juvenile patients with a history of deep vein thrombosis, confirmed by phlebography.

These numbers include only thrombosis patients, not the family members. Our collective covers 205 patients, 113 males and 92 females. The age of manifestation of thrombosis was below 45 years. 46 patients were investigated (2) 1 - 4 weeks after thrombosis and 159 patients after more than 4 weeks. The localisation of the thrombosis can be seen in Tab. 2.

Protein C - Biochemical and Medical Aspects

Table 1. Incidence of inherited fibrinolysis disorders in juvenile thrombosis patients
(Diagnosis: DVT - diagnosis by phlebography and clinical examination)

Plasminogen ↓	1 %	(2/197)
Abnormal plasminogen	1 %	(2/197)
AT III ↓	3 %	(6/200)
Protein C ↓	6 %	(8/132)
Plasminogen activator release ↓	15 %	(17/110)

Table 2. Localisation of thrombosis in juvenile patients

Deep vein thrombosis	n = 148	
Arm vein thrombosis	n = 26	
Multilocalized	n = 9	
Mesenterial v.t.	n = 1	
Central v.t.	n = 3	
Pulmonary embolism and unknown thrombosis	n = 18	
Recurrent thrombosis	n = 81	= 40 %

The distribution of risk factors is listed in Tab. 3.

Table 3. Risk factors in juvenile patients with vein thrombosis

None	n = 127
Posttraumatic/postoperative	n = 33
Ovulation inhibitors/pregnancy	n = 31
Familiar disposition	n = 10
Malignancy	n = 4

The statistical calculations were done with the Kolmagoroff-Smirnoff-test and the T-test.

One of the early published causal factors for thrombogenesis is deficiency of antithrombin III. The frequency of this inherited defect in thrombosis patients was estimated to be between 2 and 5 %. In our patients we established an occurrence of 3 %. One of our patients suffered from recurrent deep vein thrombosis since he was 12 years old. In the literature the incidence was compared to the frequency of hemophilia A.

In the inherited AT III deficiencies, 3 types can be differentiated. Type I: functional and antigen low; type II: functional low, antigen normal; type III: borderline or reverse type (?) - functional low, antigen lower; type II can further be classified into 3 groups (Tab. 4).

Table 4. Types of AT III deficiency

Type		
Type 1	a -	normal heparin affinity
	b -	↓ heparin affinity
Type 2	a -	↓ F Xa and thrombin inactivation abnormal heparin - AT III reaction and immunochem. structure
	b -	↓ F Xa and thrombin inactivation normal heparin - AT III reaction and immunochem. structure
	c -	isolated disorder of heparin- AT III reaction ↓ heparin affinity and/or ↓ heparin cofactor normal thrombin or F Xa inactivation normal immunochem. structure

Here the defect Xa and thrombin inactivation and the disturbed heparin AT III reaction can be distinguished. The abnormal AT III molecules were named after the towns, where these patients were first examined.

During dicoumarin treatment the AT III level can increase. This fact must be taken into consideration when analysing blood from anticoagulated patients. In acquired AT III deficiency a decreased synthesis (for example in liver cirrhosis), an increased consumption (for example in DIC) and an enhanced loss (for example in nephrotic syndrome) can be differentiated.

Another factor involved in thrombogenesis is the molecule plasminogen. The decreased content of plasminogen in blood or the abnormal molecule can be responsible for the incidence of thrombosis. We found a decreased level of plasminogen, measured as antigen and as functional activity as well, in 2 families with high incidence of thrombosis.

An abnormal molecule of plasminogen was described by Aoki et al. in Japan (1), by Wohl et al. in Chicago (8), by Soria et al. in Paris (7) and by our group (6). In these patients the lower proportion between the functional activity and the antigen is evident. The plasmin generation rate by addition of streptokinase (Fig. 1) and of urokinase (Fig. 2) is decreased.

Tab. 5 demonstrates the findings of Robbins et al. and of our patients.

We suggest that the discrepancy between biological and antigenic determination plays an important role in establishing the cinical diagnosis. Tab. 6 demonstrates the published data regarding the discrepancy between biological and antigenic determinations of plasminogen.

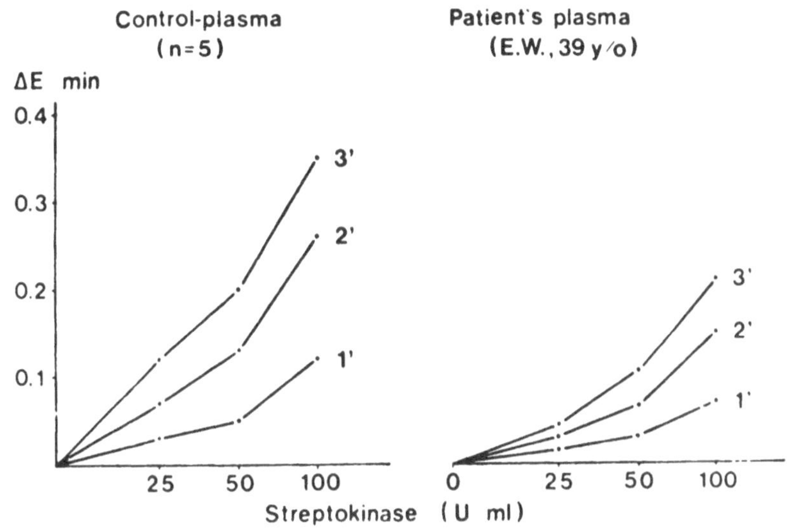

Fig. 1. Plasmin generation after addition of streptokinase.

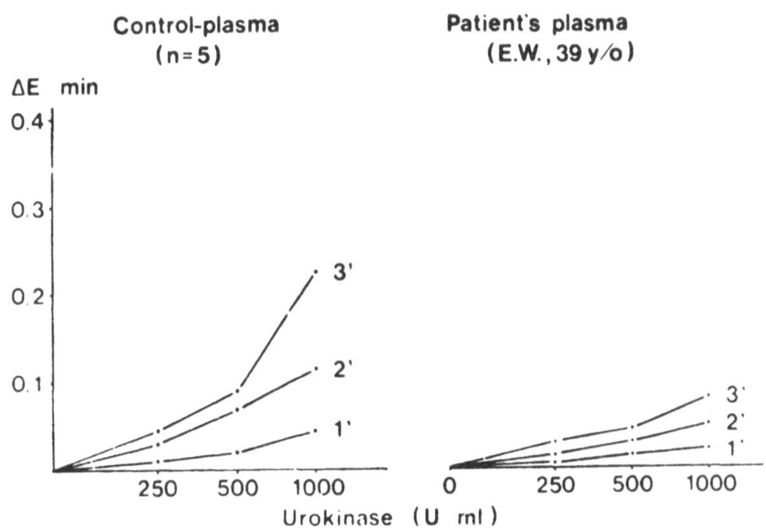

Fig. 2. Plasmin generation after addition of urokinase (U/ml)

Table 5. Comparison of plasminogen (functional and antigen concentration) and plasmin generation rate in plasma of patients with abnormal plasminogen variants.

	Plasminogen		Plasmin	
	Functional	Antigen	generation rate	
	mg/dl		UK %	SK %
Normal range	16.7-23.8	17.4-21.6	0.25-0.47	5.3-9.7
Chicago I	7.6	18.7	0.1	2.7
Frankfurt I (patient)	7.5	15.6	0.26	0.51
Frankfurt I (mother)	6.7	14.3	0.28	1.33
Frankfurt I (sister)	17.6	20.8	0.43	1.59
Frankfurt I (nephew)	16.4	19.0	0.42	0.72
Frankfurt II	12.0	18.0	0.54	0.16

Table 6. Discrepancy between biological and antigenic determinations of plasminogen

0.2	Tokio	Kazama et al.	1981/3
0.41	Chicago I	Wohl et al.	1979/8
0.5	Tochigi	Aoki et al.	1978/1
0.5	Paris	Soria et al.	1982/7
0.81	Chicago III	Wohl et al.	1982/9
0.88	Chicago II	Wohl et al.	1979/8
0.48	Frankfurt I		
0.6	Frankfurt II		

A quotient below 1 in a patient with typical clinical symptoms and no other abnormal coagulation and fibrinolysis values should lead to further investigations with performance of plasmin generation rate and isolation and purification of the abnormal plasminogen molecule. Considerations for differential diagnosis of abnormal plasminogen molecule are hypoplasminogenemia and urokinase and streptokinase inhibitors. Robbins and coworkers distinguished 3 classes (Tab. 7) of plasminogen variants.

Table 7. Classes of plasminogen variants

Class	Abnormal	Normal
A Tochigi	active center defect, charge mutation difference	normal cleavage of the Arg_{560}-Val peptide bond
B Chicago I Chicago II	K_m of activation defect with impaired activator binding	normal cleavage of the Arg_{560}-Val peptide bond
C Chicago III	K_m of activation defect with impaired activator binding K_{cat} of activation defect with impaired Arg_{560}-Val peptide bond cleavage	

Wohl et al. 1982

They drew the conclusion from their data that in abnormal zymogens the binding properties of activators as well as the active sites formed in the equimolar zymogen streptokinase complexes were impaired.

Besides the concentration of plasminogen and the behaviour of the plasmin generation rate, the reduction of plasminogen activators was also discussed as a cause for venous thrombosis. The plasminogen activator level can be measured after venous occlusion for 20 minutes or after DDAVP infusion on fibrin plates. An impairment can be caused by either a decreased release from the venous endothelium or by a reduced content in endothelium or by occurrence of a specific inhibitor against TPA.

The published frequency in thrombosis patients varies widely, from 73 % in the collective of Nilsson et al. 1983 (5) to 33 % in the Vienna patients (4). In our patients we also found a high incidence: 15 % with decreased release reaction in repeated investigations. In these patients we found a striking high incidence of recurrent thrombosis.

Thromboembolic diseases in patients with dysfibrinogenemia were published more often in the last years (fibrinogen "Baltimore", "Haifa", "New York", "Oslo" and "Paris").

Another factor in thrombophilia is protein C. The incidence of protein C deficiency was estimated between 1 - 2 % in thrombosis patients; probably the frequency is much higher. In our juvenile thrombosis patients, we found a percentage of approximately 6 %. These patients had no liver disease or DIC. All plasmas were tested by the Laurell technique and Elisa in repeated investigations. During stable anticoagulant therapy (Tab. 8), the protein C levels vary between 30 - 50 % in our patients with vitium cordis.

Table 8. Investigation of a stable dicoumarin collective (n = 29, TPT: 17.7 %)
- Protein C -

	mean value (%)	SD	range	VC
Protein C, Elpho	43.18	14.9	28 - 58	34.55
Protein C, Elisa	36.17	14.3	21 - 50	39.78
II (antigen)	47.48	10.1	37 - 57	21.35
X (antigen)	44.86	7.38	37 - 52	16.46
PC AG/II AG	0.9	0.26	0.64 - 1.17	29.61
PC AG/X AG	0.96	0.27	0.68 - 1.23	28.8

In anticoagulant patients with inherited protein C deficiency the levels are lower.

Tab. 9 compares the behaviour of AT III and protein C.

Up to now we found no combined defect. Especially our patients with plasminogen deficiency, with abnormal plasminogen molecule, with antithrombin III deficiency and with reduced plasminogen activator release had normal protein C levels. Furthermore our 8 patients with protein C deficiency were investigated for plasminogen levels, abnormal plasminogen molecule, antithrombin III deficiency and plasminogen activator release. No defect or other deficiency could be demonstrated.

Summarizing, in our juvenile thrombosis patients (n = 200 investigated patients) we found a percentage of 25 % inherited coagulation and fibrinolysis defects. These findings are important because of the therapeutic consequence of a life long anticoagulant prophylactic treatment in these patients.

Table 9. Comparison of AT III and protein C

	AT III	Protein C
plasma protein	alpha$_2$-globulin	zymogen
molecular weight	65,000 daltons	62,000 daltons
synthesis site	liver	liver
plasma concentr. antigen activity	 22 - 39 mg/dl 80 - 120 %	 0.99 $\pm$ 0.17 U/ml 0.94 $\pm$ 0.17 U/ml
half-life	2.8 $\pm$ 0.3 days	5 - 7 hours
biological characteristics	inhibition of serine proteases, inhibition of thrombin, cof. of heparin efficacy	serine protease inactivator of the activ. F VIII and F V
behaviour during oral anticoag. therapy	increase	decrease
therapy of deficiency	AT III concentrate oral anticoagulants	oral anticoagulants

References

1. Aoki, N., Moroi, M., Sakata, Y., Yoshida, N.: J. Clin. Invest. 61, 1186-1195 (1978)

2. Hach, V., Walter-Fincke, R., Maas, C., Heyland, H., Scharrer, I.: Poster anläßlich des 3. Kongresses für Thrombose und Blutstillung, Bern 1984, Abstract band, P.2.3, p. 56 (1984)

3. Kazama, M., Tahara, C., Suzuki, Z., Gohchi, K., Abe, T.: Thromb. Res. 21, 517-522 (1981)

4. Korninger, C., Lechner, K., Niessner, H., Gössinger, H., Kundi, M.: Thromb. Haemostas. 52, 127-130 (1984)

5. Nilsson, J.M., Tengborn, L.: In: Clinical Aspects of Fibrinolysis and Thrombolysis. Sysjysk Universitetsvorlag, Esbjerg, 19 (1983)

6. Scharrer, I., Robbins, K.C., Wohl, R.C., Hach, V., Maas, C.: Thromb. Haemostas. 50, 257 (1983)

7. Soria, J., Soria, C., Bertrand, O., Lefur, M.A., Drouet, L., Baudony, Y., Dupuy, E.: Hemostasis 11, 1 (1982)

8. Wohl, R.C., Summaria, L., Robbins, K.C.: J. Biol. Chem. 254, 9063-9069 (1979)

9. Wohl, R.C., Summaria, L., Robbins, K.C., Chediak, J., Rosenfeld, S.: Thromb. Haemostas. 48, 146-152 (1982)

PROTEIN C CONCENTRATION AFTER ELECTIVE HIP SURGERY

Theodor H. Schöndorf and Irene Witt
Zentrum für Innere Medizin der Justus-Liebig-Universität
D-6300 Giessen
Biochemisches Labor der Universitäts-Kinderklinik
D-7800 Freiburg

Introduction

The high incidence of deep venous thrombosis and pulmonary embolism after emergency or elective hip surgery is well known (1). In previous investigations (2, 3) a significant decrease of antithrombin III activity was found, and in 55 - 60 % of the patients an elevation of fibrin monomers up to the 3rd postoperative day could also be demonstrated.
The occurrence of soluble fibrin monomers results from the increased intravascular thrombin formation due to activation of the coagulation system by operative tissue injury. Antithrombin III decreases because it is consumed by complex formation with activated coagulation factors.
Mannucci and Vigano (4) reported an acquired deficiency of protein C in patients after different kinds of surgery. The protein C concentration which was measured immunologically decreased by 20 % compared to the preoperative values on the first and second postoperative day. The authors interpreted the decrease as a result of consumption by increased thrombin formation due to in vivo activation of blood coagulation, similar as has been demonstrated for antithrombin III.
The aim of our study was to evaluate if a protein C decrease might contribute to the high risk of thromboembolism after hip surgery.

Methods

Protein C antigen (PCA) concentration was measured by an Elisa kit according to the recommendations of the manufacturer (Boehringer Mannheim, FRG).
Antithrombin III activity was determined by a chromogenic substrate assay (Boehringer Mannheim, FRG). Factor VII activity was also measured with a chromogenic test system (4).
Fibrinogen was measured functionally with the clotting test according to Clauss and α_2-macroglobulin immunologically with the Immmunchemical System (Beckman Instruments, München, FRG).

Patients

The 20 patients (12 females, 8 males) whose plasma was analyzed were between 48 and 78 years old ($\bar{x}$ = 64.2). They were in good clinical condition except for their degenerative coxarthritis.

Results

After elective hip replacement operations we observed a reduction of the coagulation inhibitors protein C, antithrombin III and α_2-macroglobulin with lowest values on the first postoperative day (Fig. 1). During the following days the inhibitors increased to preoperative values.
Fig. 2 shows the decrease of protein C together with the decline of the procoagulatory factor VII, which both are dependent of vitamin K. In contrast fibrinogen increased considerably because it is an acute phase reactant.
During operations the hematocrit is diminished by blood loss and subsequent volume replacement. In the investigated pa-

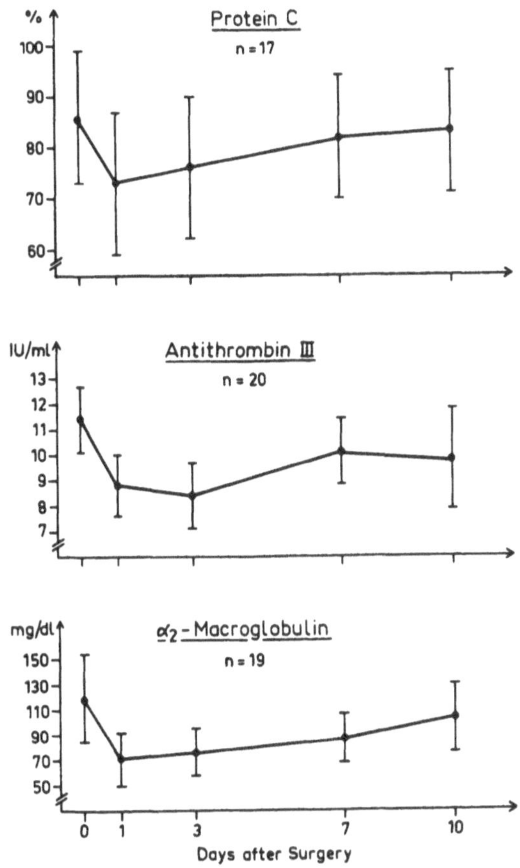

Fig. 1. Decrease of plasma inhibitors after elective hip surgery.

tients the average decrease of hematocrit on the first post-operative day was nearly 20 %.

In comparison protein C decreased by 15 % and the activities of antithrombin III and α_2-macroglobulin by 30 % and 40 % respectively. Thus it is assumed that the decrease of protein C is a result of blood dilution since it does not exceed the reduction of hematocrit. The more pronounced decline of antithrombin III and α_2-macroglobulin is suggested to be

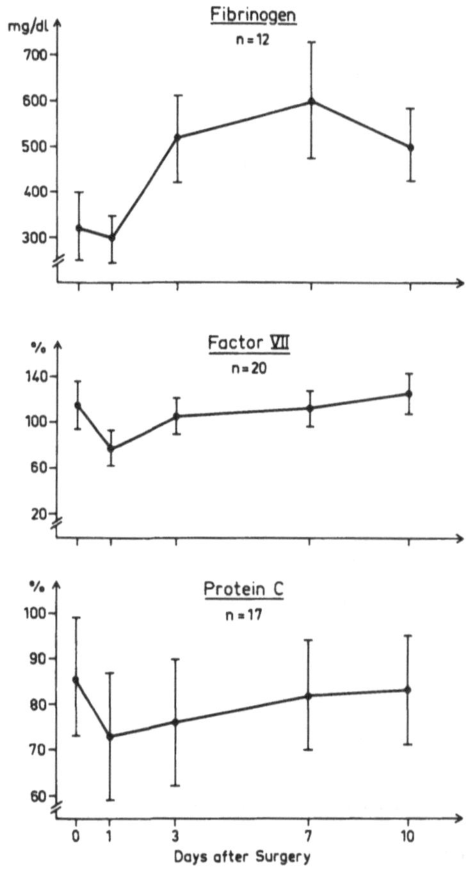

Fig. 2. Changes of coagulation proteins after elective hip surgery.

not only caused by dilution but also as a consequence of consumption by activated coagulation factors.

An additional indication for the activation of coagulation is the marked decrease of the coagulation factor VII.

The results of our study are in accordance with the observation of Mannucci that protein C decreased after surgery, but we cannot share the opinion that it is consumed during activation of blood coagulation.

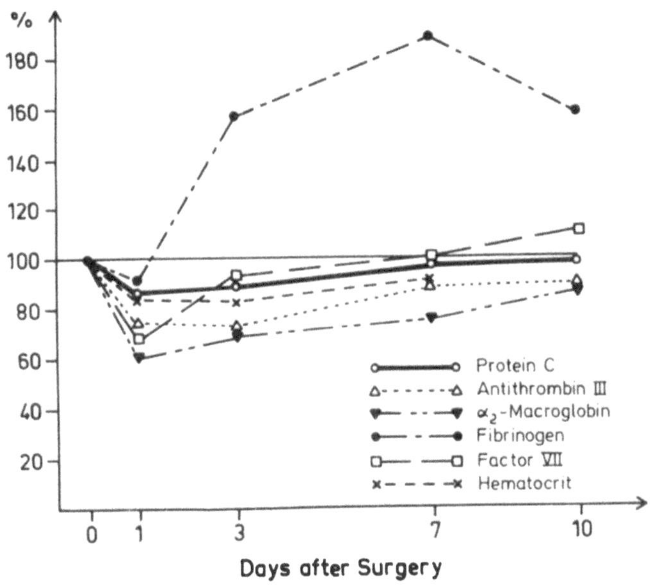

Fig. 3. Relative changes of hematocrit and coagulation proteins after elective hip surgery

References

1. Bergqvist, D.: Postoperative Thromboembolism, Frequency, Etiology, Prophylaxis. Springer-Verlag, Berlin-Heidelberg-New York 1983

2. Schöndorf, T.H., Oehler, G.: Thromb. Haemostas. 50, 50 (1983)

3. Mannucci, P.M., Vigano, S.: Lancet ii, 463-467 (1982)

4. Witt, I., Krause, P.: 29. Jahrestagung der DAB/GTH, 1985, Abstract HP 7.7.4

PROTEIN C ACTIVITY AND CONCENTRATION IN PATIENTS WITH THROMBO-EMBOLIC DISEASE AND IN PATIENTS TREATED WITH PHENPROCOUMON

G. Müller-Berghaus, W. Thiel, K.T. Preissner and G. Oehler
Klinische Forschungsgruppe für Blutgerinnung und Thrombose,
Max-Planck-Gesellschaft
Zentrum für Innere Medizin der Justus-Liebig-Universität,
D-6300 Giessen

Introduction

Several functional assays of protein C have been described. When measuring plasma protein C activity, some problems arise from its low concentration, the endogenous inhibitors of protein C and the unspecificity of the chromogenic substrates used for measuring protein C activity. In the present paper we describe a functional assay which is similar to the one published by Sala et al. (1).

Method

The principle of the functional assay is outlined in Tab. 1. Carboxylated protein C is recovered from 1 ml of plasma by adsorption to barium citrate and consecutive elution with EDTA. Thereafter, EDTA is removed by precipitation of the protein with PEG. Protein C is activated with thrombin and the cofactor thrombomodulin which was derived from rabbit lung homogenate. After activating protein C, thrombin and other enzymes are inactivated by antithrombin III and hirudin, and the protein C activity within the reaction mixture is measured using the chromogenic substrate S-2266.

Protein C - Biochemical and Medical Aspects

Table 1. Principle of protein C activity assay.
(From Thiel et al. (4))

1.	Adsorption with barium citrate
2.	Elution with EDTA
3.	Activation of protein C with thrombin, thrombomodulin and Ca^{++}
4.	Inactivation of thrombin and other enzymes by antithrombin III and hirudin
5.	Measurement of activated protein C with a chromogenic substrate

Most of the investigators use antithrombin III and heparin for inhibiting thrombin in the reaction mixture. Tab. 2 gives the spectrophotometric values if barium citrate eluate was mixed with thrombin, thrombomodulin, antithrombin III and/or heparin. Tab. 2 indicates that the presence of heparin alone or together with antithrombin III gives already an adsorption independently on the presence or absence of protein C. Thus, heparin together with other plasma constituents activates the chromogenic substrate. For this reason, heparin was omitted in the assay procedure as used in this study and substituted by hirudin.

Table 2. Spectrophotometric values of plasma protein C.
(From Thiel et al. (4)).

	Barium citrate eluate mixed with				
Thrombin	Tb	Tb	-	-	-
Thrombomodulin	TM	-	-	-	-
Antithrombin III (3 U/ml)	AT III	-	AT III	AT III	-
Heparin (0.03 mg/ml)	Hep	-	Hep	-	Hep
mA/min	180	40	185	10	130

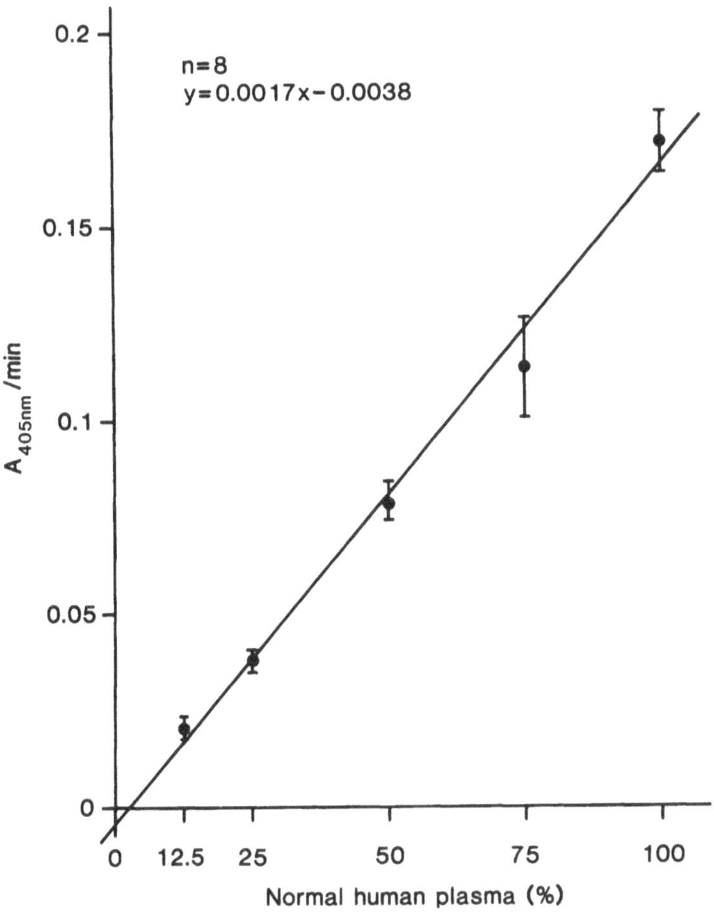

Fig. 1. Calibration curve for the determination of protein C activity. (From Thiel et al. (4)).

Results

Fig. 1 represents the calibration curve for the determination of protein C activity. Linearity of the curve was observed if human plasma was diluted from 100 to 12.5 % of normal. The assay precision for protein C activity is given in Tab. 3. The coefficient of variation within days amounted to 5.92 % and that from day to day to 7.51 %.

Table 3. Assay precision for protein C activity (n = 12). (From Thiel et al. (4)).

	Mean $\pm$ SD (U/ml)	CV %
within day	0.90 $\pm$ 0.05	5.92
from day to day	0.96 $\pm$ 0.07	7.51

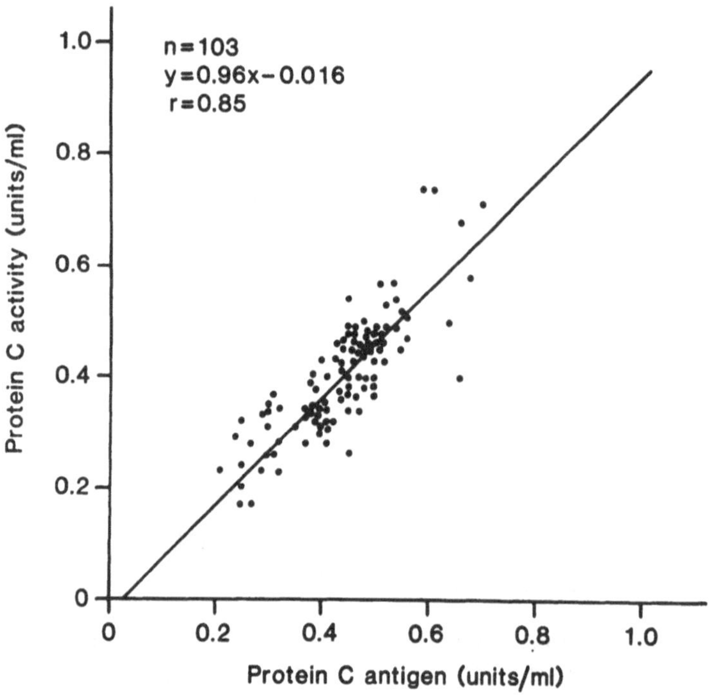

Fig. 2. Correlation of protein C activity and protein C antigen determined in 103 phenprocoumon-treated patients. (From Thiel et al. (4)).

The functional assay was compared with the immunological determination of protein C using a commercially available ELISA kit from Boehringer Mannheim. The determinations were performed on microtitre plates.

Protein C activity and concentration were determined in plasma samples of 53 healthy volunteers. The protein C activity ranged from 0.63 to 1.42 U/ml and the protein C antigen of the same samples from 0.62 to 1.30 U/ml.
Protein C activity and concentration were determined in 103 phenprocoumon-treated patients. Protein C activity and concentration correlated to each other with a correlation coefficient of 0.85 (Fig. 2).

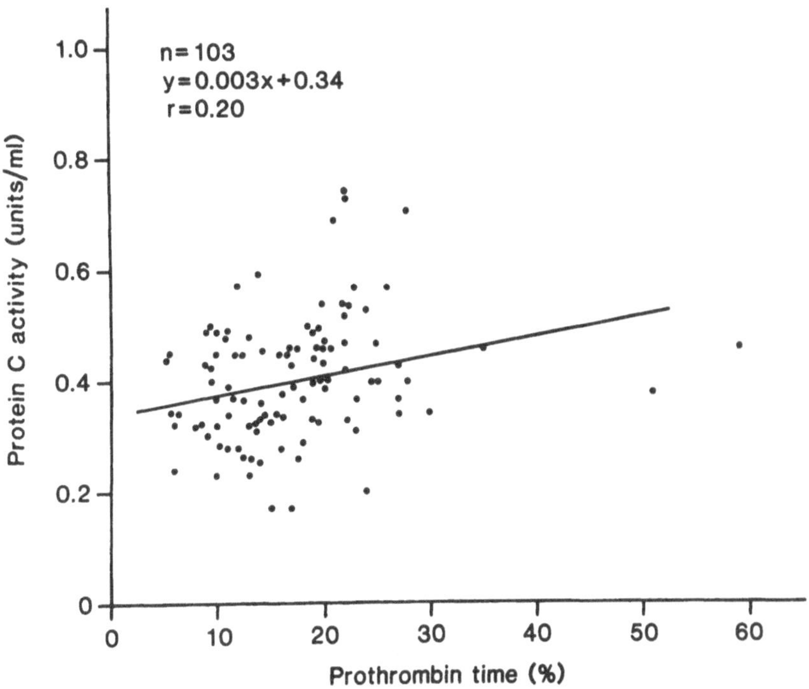

Fig. 3. Missing correlation between protein C activity and the prothrombin time in 103 phenprocoumon-treated patients. [From Thiel et al. (4)].

In contrast to other investigators using a different assay system (2), our data did not demonstrate higher protein C antigen values than protein C activities in phenprocoumon-treated patients. If the protein C activity was compared with

Table 4. Protein C in healthy individuals and in patients' plasma. [From Thiel et al. (4)].

	N	Protein C activity (U/ml) Mean ± SD	Protein C antigen (U/ml) Mean ± SD
Normals	53	0.93 ± 0.23	0.91 ± 0.13
Phenprocoumon-treated	103	0.40 ± 0.11	0.43 ± 0.10
Liver cirrhosis	24	0.31 ± 0.15	0.34 ± 0.15
DIC	5	0.15 ± 0.06	0.33 ± 0.19

the prothrombin time, no correlation was found with the 103 patients studied (Fig. 3). These data indicate that the prothrombin time does not give any information about protein C activity and agrees with the observation that the degradation of factor Va by activated protein C does not play any decisive role in the determination of the prothrombin time. Because of its short biological half-life time, protein C activity may more pronouncedly fluctuate with patients under phenprocoumon treatment than factor II and factor X activities.

In five patients with acute disseminated intravascular coagulation low protein C activity and low protein C antigen were determined (Tab. 4). These data agree with the observation that the generation of intravascular thrombin may activate protein C on the surface of endothelial cells causing a consumption of protein C similar to the consumption of other coagulation factors in the course of DIC. Patients with disseminated intravascular coagulation demonstrated a ratio of protein C activity to protein C antigen below 1.0, indicating that some of the circulating protein C might be inactivated by an inhibitor against activated protein C (3). Thus, protein C activity may be lower than protein C antigen because

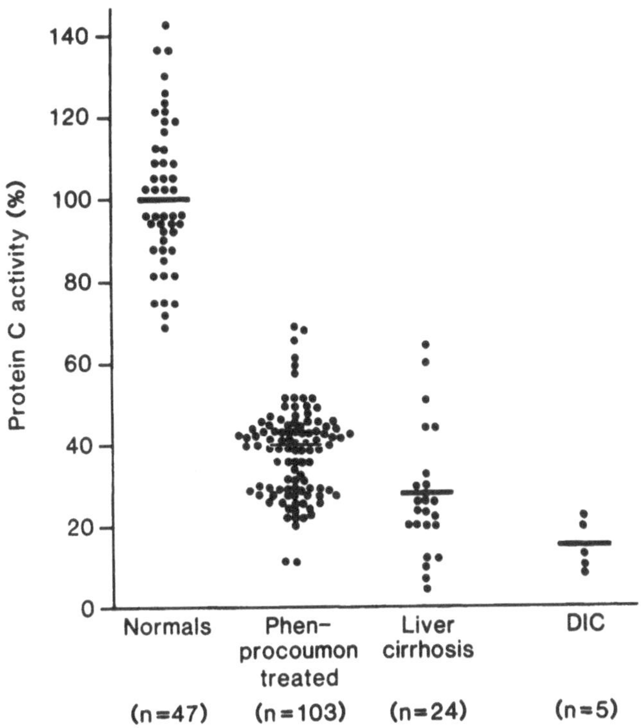

Fig. 4. Different mean protein C activity levels in phenprocoumon-treated patients and in patients with liver cirrhosis or disseminated intravascular coagulation (DIC). (From Thiel et al. (4)).

of complex formation between protein C and its inhibitor in plasma.

Fig. 4 summarizes additional data obtained with patients with liver cirrhosis. In these patients protein C activity as well as protein C antigen were very low because of impaired protein synthesis. In some patients the ratio of protein C activity to protein C antigen was far below 1.0 indicating the synthesis of a functional deficient protein or the formation of a complex between protein C a (activated protein C) and protein C a inhibitor.

The data presented indicate the importance and need of functional assays of protein C. Such an assay is necessary for diagnosing type II protein C deficiency as well as studying protein C in regulating intravascular coagulation. So far, protein C activity assays are too complicated, time-consuming and expensive to be used for a wide range.

References

1. Sala, N., Owen, W.G., Collen, D.: Blood 63, 671-675 (1984).

2. Bertina, R.M., Broekmans, A.W., Krommenhoek-van Es, C., van Wijngaarden A.: Thromb. Haemostas. 51, 1-5 (1984).

3. Suzuki, K., Nishioka, J., Hashimoto, S.: J. Biol. Chem. 258, 163-168 (1983).

4. Thiel, W., Preissner, K.T., G. Müller-Berghaus: Submitted for publication.

PROTEIN C UNDER LONG-TERM DICOUMAROL TREATMENT

Peter Hellstern, Gerhard von Blohn, Chieko Miyashita, Michael Köhler, Peter Scheffler, Ernst Wenzel
Abteilung für Klinische Hämostaseologie und Transfusionsmedizin, Universität des Saarlandes, D-6650 Homburg-Saar

Introduction

Vitamin K deficiency and oral anticoagulation (OA) lead to a decrease of protein C (PC) synthesis. The purpose of this paper was to investigate the plasma PC levels during the initial phase and during the stable phase of OA.

Methods

A commercially available micro ELISA (Boehringer Mannheim) was used to measure PC concentrations in citrated platelet poor plasma (cPPP). Micro ELISA plates were utilized. The laboratory monitoring of patients on OA was performed with Calcium-Thromboplastin and the Hepato-Quick reagent (Boehringer Mannheim). In addition, various other thromboplastins were applied.
The normal range of PC concentration was obtained from 100 blood donors (50 males, 50 females). Calibration curves were obtained using a citrated pooled plasma from 30 healthy individuals. Five patients were monitored during the initial phase of OA, and 20 patients in the stable phase of OA. Thirteen males and seven females were monitored 6 times at 2 to 3-week intervals (Tab. 1).

Protein C - Biochemical and Medical Aspects

Table 1. Characteristics of the patients under stable oral anticoagulant treatment

Patients: 13 males, 7 females
stable anticoagulation:
Ca-Thromboplastin: $\bar{x}$ = 25.5% (SD = 6.3 %)
HQ: $\bar{x}$ = 14.4 % (SD = 4.6 %)
monitored 6 times in 2 to 3-week intervals
12 patients: no change in dicoumarol dosage
8 patients: slight changes of less than 15 % of individual dosage
no bleeding or thromboembolism

Results

The normal range of PC was found to be 64 - 150 %. In the initial phase of OA the PC levels decreased in close correlation to the Quick and Hepato-Quick values (Fig. 1). Once the therapeutic range was attained, PC levels were in the range of 31 - 61 %.

The overall mean of PC values of patients in the stable phase of OA was 50 % (SD = 12.6 %). The intraindividual and the interindividual variations of PC are comparable with the values of variation of thromboplastin times (Tab. 2).

There was a poor correlation between the Calcium-Thromboplastin time values and the PC values in patients during the stable phase of OA (Fig. 2).

Five other thromboplastins with a proven different sensitivity for decrease of F II, VII, and X were correlated with PC levels, but correlation was also weak (Tab. 3).

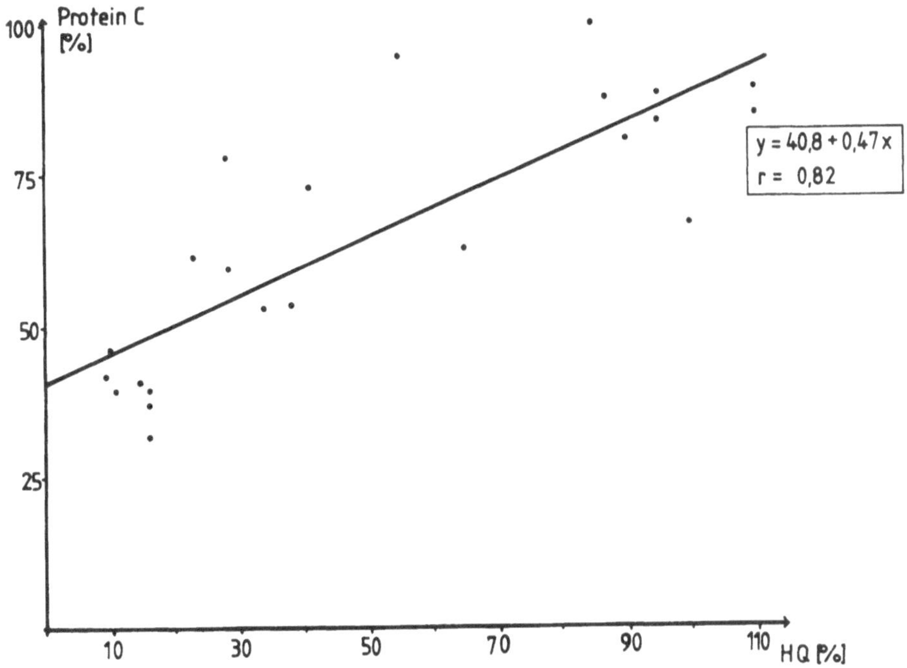

Fig. 1. Correlation between Hepato-Quick (HQ) and PC during the initial phase of oral anticoagulation

Table 2. Interindividual and intraindividual variation of PC, Calcium-Thromboplastin time, and Hepato-Quick values in patients under stable oral anticoagulant treatment.
This should not be mistaken for between-run and within-run precision.

Protein C	CV intra = 22.3 %
	CV inter = 40.0 %
Ca Thromboplastin	CV intra = 21.3 %
	CV inter = 43.6 %
Hepato-Quick	CV intra = 26.8 %
	CV inter = 51.4 %

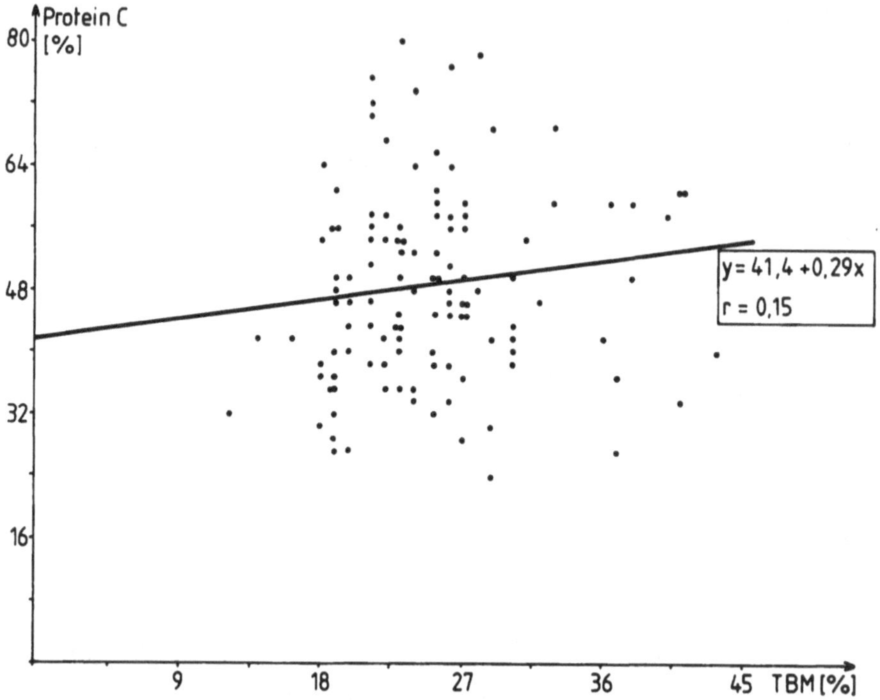

Fig. 2. Correlation between Calcium-Thromboplastin time (Boehringer Mannheim) and PC during the stable phase of oral anticoagulation

Table 3. Correlation (r) between protein C and prothrombin time (n = 120)

Ca-Thromboplastin:	0.15
Hepato-Quick:	0.20
Simplastin:	0.03
Thromborel:	0.10
Normotest:	0.07
Thrombotest:	0.07

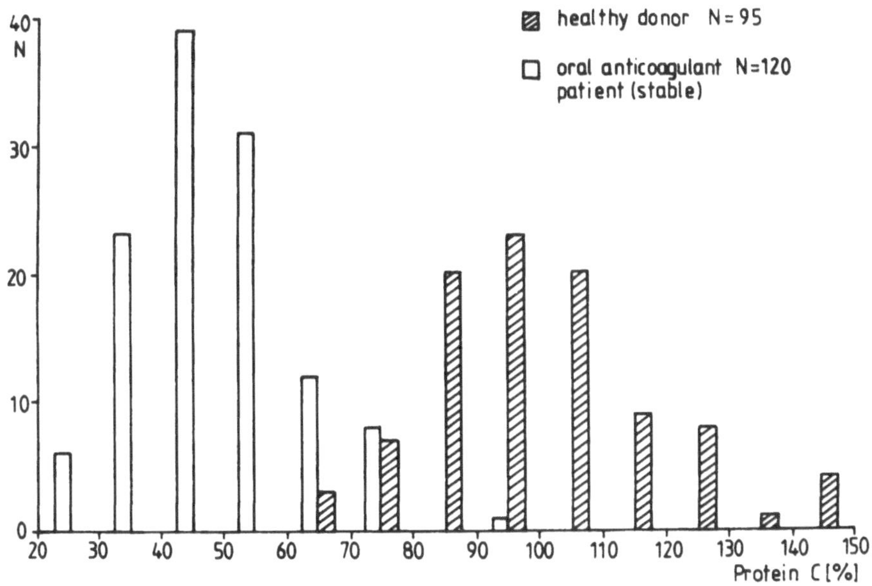

Fig. 3. Normal range of PC and range of PC values under stable oral anticoagulation

Conclusions

In accordance with other authors we found a rapid decrease of PC in close correlation to the Hepato-Quick values during the initial phase of OA. During the stable phase of OA the PC levels are decreased to 50 % of the mean of healthy subjects. There is some overlap between the normal range and the values under OA (Fig. 3). This indicates a somewhat unsatisfactory sensitivity and efficacy of PC concentration measurements for the discrimination of vitamin K deficiency or vitamin K antagonists. The interindividual and intraindividual variation of the values is relatively high for PC as well as the coagulation assays. However, the between-run and within-run precisions of PC determination using the micro ELISA technique do not exceed 10 %, and this is satisfactory for the measurement of protein concentrations.

PROTEIN C UNDER HEPARIN TREATMENT

Ulrich Schmitz-Huebner
Medizinische Klinik und Poliklinik der Westfälischen Wilhelms-Universität, D-4400 Münster

Introduction

Heparin and its derivatives have several effects on the clotting system which up till now have only partially been elucidated (1). The mechanisms responsible for the antithrombotic activity of this substance are still poorly understood (1, 2). Protein C has been shown to possess the capacity to inhibit F Va and F VIIIa and to potentiate the fibrinolytic activity in the blood of some species. It has been suggested that heparin might interact with the activation and inhibition mechanisms of protein C (3, 4).
In the present experiments protein C antigen (PC Ag) levels were determined before and after administration of different types of heparin to healthy volunteers as a preliminary attempt to evaluate the effect of heparin treatment on protein C under clinical conditions.

Materials and methods

1. <u>Heparin preparations.</u> Unfractionated porcine sodium mucosal heparin (UFH), mean molecular weight 12,500. Low molecular weight heparin (LMWH) prepared from UFH by ultrafiltration (Braun Melsungen AG, 3508 Melsungen, FRG), mean molecular weight 9,500.

Protein C - Biochemical and Medical Aspects

2. Subjects. The UFH and LMWH preparations were administered by i) subcutaneous injections to six healthy individuals and by ii) intravenous injections to eight healthy individuals belonging to both sexes, aged 20 - 28. Each subject received both types of heparin preparations in a cross-over trial after informed consent had been given. Intervals of at least one week were observed between the treatments.

3. Plasma. Nine parts of venous blood were mixed with one part of 3.8 % (w/v) sodium citrate solution and immediately centrifuged at 4°C and 2,500 g for 40 min. Normal plasma pools destined for the establishment of standard curves were obtained from at least 20 healthy donors and processed in the same manner. Plasma samples were snap-frozen and stored at - 30°C.

4. Heparin assays. The anti-Xa activity and the anti-IIa activity induced by heparin were determined by chromogenic substrate assays (5) making use of the synthetic substrates S-2222 and S-2238 by KABI AG, Munich.

5. Protein C assay. The determination of protein C antigen (PC Ag) in plasma was performed by means of an ELISA kit according to the prescriptions of the manufacturer (Boehringer Mannheim, FRG). Coefficients of variation were as follows: intra-assay (n = 15) 3.49 %, inter-assay (n = 7) 6.69 %. The content of PC Ag in the plasma examined was within the normal range.

6. Statistical analysis. Paired samples were compared with the Wilcoxon test (6). The analysis of the PC Ag distribution in normal subjects was performed with the Kolmogoff test.

Results

1. Distribution of PC Ag in healthy individuals.
   Fig. 1 shows the results obtained by measuring PC Ag in the plasma of 100 healthy subjects, aged 20 - 40 years. The mean value was 90.6 % with a standard deviation of 13.3 %. The results are normally distributed between the extremes of 60 and 125 %.

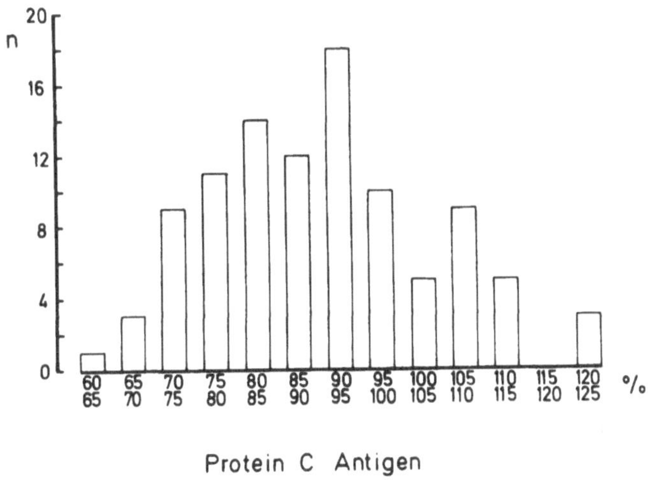

Fig. 1. Results of PC Ag determinations in 100 healthy individuals.

2. Intravenous administration of heparin.
   As shown in Fig. 2 the i.v. administration of 2,000 IU of ordinary heparin causes insignificant changes of the PC Ag concentrations, although quite considerable levels of heparin activity are reached for at least 60 min. More pronounced effects are to be seen after i.v. injection of the same dose of the LMWH fraction (Fig. 3). After an initial decrease PC Ag values show a marked but short-termed increase and return to the preinjection level 60 min after injection. When the effects of the two heparin prepara-

tions on PC Ag are compared, a significant difference in the values obtained 30 min after injection is observed.

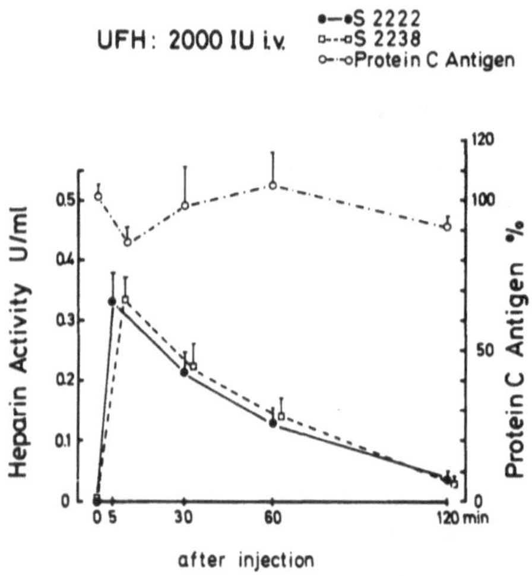

Fig. 2. PC Ag levels and heparin activities before and after i.v. administration of unfractionated heparin. Mean values ± SEM, n = 8.

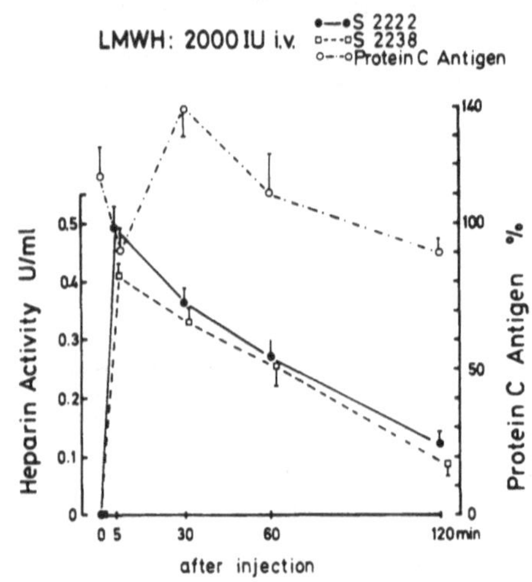

Fig. 3. PC Ag levels and heparin activities before and after i.v. administration of the LMWH fraction. Mean values ± SEM, n = 8.

3. Subcutaneous administration of heparin.
Similar doses of the two heparin preparations administered subcutaneously b.i.d. during 3 days did not induce significant changes of the PC Ag concentrations. PC Ag was determined before and 120 min after s.c. injection on the first and third day of the trial (Fig. 4).

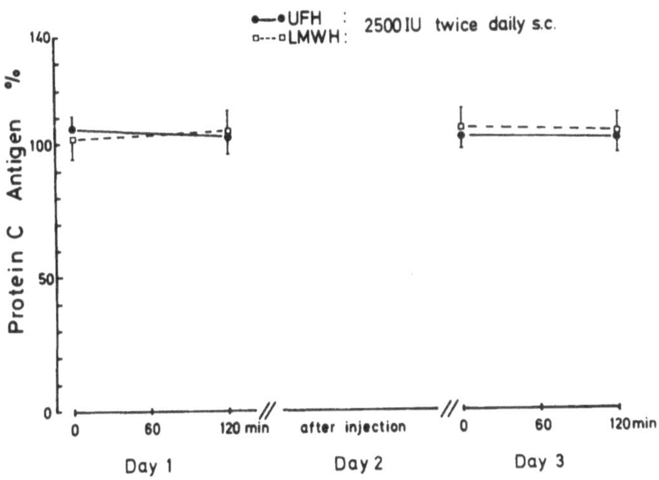

Fig. 4. PC Ag levels before and after s.c. administration of unfractionated heparin or the LMWH fraction during three days b.i.d. Mean values ± SEM, n = 6.

Discussion

The results presented indicate that the immunologic assay used in this study represents a precise and reproducible method to determine protein C concentrations in plasma. A normal distribution of the values measured in 100 healthy individuals could be documented. The mean value of 90.6 % was rather low and this points to the fact that pool plasma destined for the establishment of a protein C reference curve should be sampled from a larger group of healthy donors.

Ordinary heparin at the doses employed in this study had no influence on protein C concentration. However, the LMWH fraction when administered intravenously seemed to have a temporary effect. While it at present appears uncertain that this effect has any relation with the characteristic properties of low molecular weight heparin, further studies employing adequate functional assays for protein C should clarify the nature of the observed alterations.

References

1. Björk, I., Lindahl, U.: Mol. Cell. Biochem. 48, 161-182 (1982).

2. Witt, I.: Heparin. New Biochemical and Medical Aspects, Walter de Gruyter, Berlin . New York 1982.

3. Esmon, C.T., Harris, K.W., Esmon, N.L., Nixon, R.N., De Bault, L.E.: Thromb. Haemostas. 50, 419 (1983).

4. Suzuki, K., Nishioka, J., Hashimoto, S.: J. Biol. Chem. 258, 163-168 (1983).

5. Schmitz-Huebner, U., Balleisen, L., Asbeck, F., van de Loo, J.: Thromb. Haemostas. 46, 612-616 (1981).

6. Sachs, L.: Angewandte Statistik, Springer, Berlin . Heidelberg . New York 1978

AUTHOR INDEX

Subject Index

Zeitfracht Medien GmbH
Ferdinand-Jühlke-Straße 7
99095 Erfurt, Deutschland
produktsicherheit@kolibri360.de